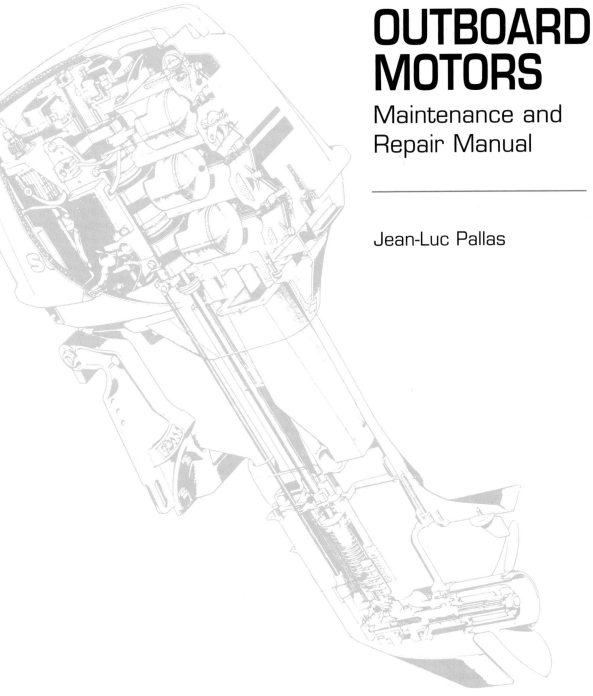

OUTBOARD MOTORS
Maintenance and Repair Manual

Jean-Luc Pallas

OTHER TITLES OF INTEREST

MARINE ELECTRICAL & ELECTRONICS BIBLE, 3rd ed.
John C. Payne

"A bible this really is...the clarity and attention to detail make this an ideal reference book that every professional and serious amateur fitter should have to hand."
Cruising

"All in all, this book makes an essential reference manual for both the uninitiated and the expert."
Yachting Monthly

"...a concise, useful, and thoroughly practical guide.... It's a 'must-have-on-board' book."
Sailing & Yachting (SA)

METAL CORROSION IN BOATS, 3rd ed.
Nigel Warren

Knowledge of metal corrosion and its prevention are vital for the safety and integrity of a boat. The third edition of this unique book is for boat owner and boat builder alike, giving advice on how to prevent or at least reduce the corrosive effects of seawater.

"One of the most coveted guides...an excellent book"
Powerboats Reports

SAILPOWER
Peter Nielsen

Any and all seafarers, whether racer or cruiser, amateur or expert, will benefit from Nielsen's down-to-earth practical advice and superb illustrations that demonstrate exactly how to achieve a perfect set-up to get the most out of your sails.

"Sailpower is a superb guide to honing one's sailing skills and highly recommended for anyone considering the hobby or simply wanting to know more about what it is like."
Midwest Book Review

COASTAL NAVIGATION USING GPS
Frank J. Larkin

GPS has forever transformed the art of navigation, but most sailors don't know how to use it effectively. This new handbook from the author of the hugely successful *Basic Coastal Navigation* serves as an invaluable guide to this essential device. Author Frank Larkin merges basic GPS techniques with tried-and-true principles of coastal navigation, making for a truly comprehensive tutorial. Clearly written and easily understood, the book guides readers through the maze of GPS options, presenting a set of criteria for selecting a GPS unit appropriate for each sailor's individual needs.

"The book covers all aspects of coastal navigation, and is recommended for self-study and reference as well as navigation classrooms."
Celestaire

THE FISHERMAN'S ELECTRICAL MANUAL
John C. Payne

The Fisherman's Electrical Manual will enhance the skills of any small boat fisherman through a better understanding of the installation, troubleshooting, and repairs of any and all electrical systems associated with the rod and reel.

Subjects covered include: outboard electrics, trolling motors, downriggers and ion systems, trim tabs, batteries and battery charging, boat wiring, sonar and fish finders, VHF and DSC radio, GPS, charts, radar, and autopilots. Thorough, well-organized, and written with clarity and insight.

SHERIDAN HOUSE
America's Favorite Sailing Books
www.sheridanhouse.com

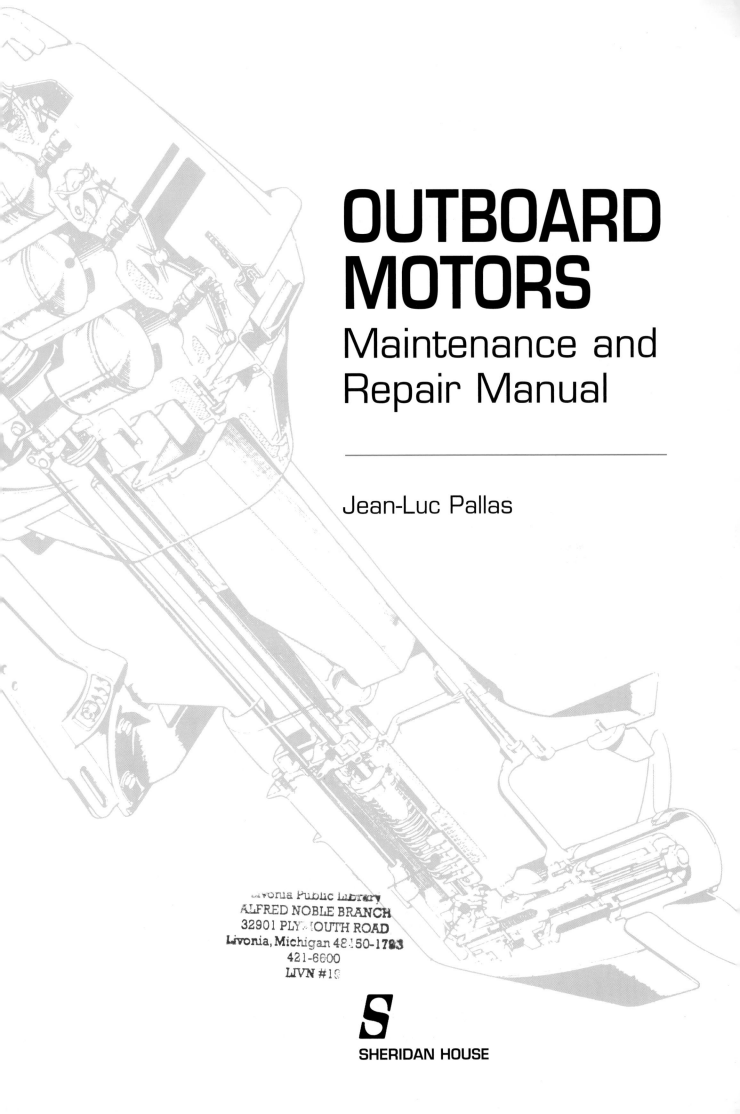

OUTBOARD MOTORS
Maintenance and Repair Manual

Jean-Luc Pallas

S

SHERIDAN HOUSE

This edition published 2006 by
Sheridan House Inc.
145 Palisade Street
Dobbs Ferry, NY 10522
www.sheridanhouse.com

First published by Editions Loisirs Nautiques 2004
under the title
Guide pratique d'entretien du moteur hors-bord
First published 2006 by Adlard Coles Nautical

While all reasonable care has been taken in the
publication of this book, the publisher takes no
responsibility for the use of the methods or products
described in the book.

Library of Congress Cataloging-in-Publication Data

Pallas, Jean-Luc.
[Guide pratique d'entretien du moteur hors-bord.
English]
Outboard motors maintenance and repair
manual/Jean-Luc Pallas.
p. cm.
Translation of Guide pratique d'entretien du moteur
hors-bord.
ISBN-13: 978-1-57409-235-6 (alk. paper)
1. Outboard motors—Maintenance and repair—
Handbooks, manuals, etc. I. Title.

VM771.P355 2006
623.87'234028—dc22
          2006016353

ISBN 1-57409-235-9
ISBN 978-1-57409-235-6

Designed by Paula McCann

Printed in Spain by GraphyCems

Contents

INTRODUCTION

AS WITH MOST TECHNICAL THINGS on a boat, the outboard engine is simple to use but can seem obscure and complex as soon as it breaks down. However, on taking a closer look and with a bit of guidance, all can become clear. That is the aim of this book. It explains engine technology in simple terms, with drawings that allow you to visualise how it works.

The best way to avoid engine breakdown is by carrying out regular maintenance. With a little guidance, a task that appears tedious and time-consuming can in fact be relatively quick and easy. The maintenance and repair worksheets in this book are designed to allow beginners as well as the mechanically knowledgeable to work on their outboard engines with confidence. All these tasks, whether maintenance or repair, are explained with precise illustrations showing the steps for each procedure. All the tasks

are coded as either *simple*, *technical* or *complex* depending on your level of skill and experience.

The book is divided into four parts. The first covers the technical elements of outboard engines. The second part contains maintenance/repair worksheets that will allow you to maintain your engine efficiently. The third section reviews the most common causes of engine breakdown. A trouble-shooting list allows you to diagnose and fix the most common outboard engine problems. Finally, the fourth part reviews the different steps to follow in one of the most important maintenance routines: winterising. Designed along the same lines as the step-by-step maintenance/repair worksheets, this chapter will show you how to winterise your engine in one short afternoon.

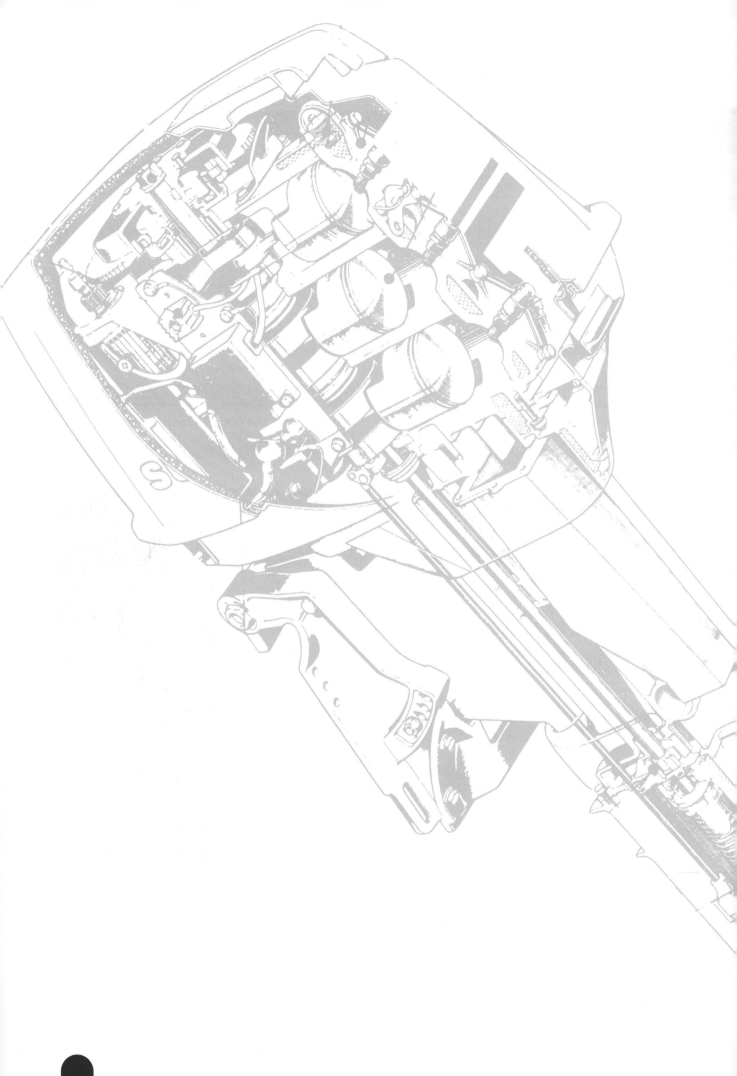

THEORY

BEFORE DOING ANY WORK on your outboard it is wise to understand some basics. Bore, flow, ignition, gearbox, valves, spark plugs... these are all terms that you need to know. In the following pages the outboard engine, whether two-stroke or four, is stripped, explained and analysed in detail so you can understand how it works. This will help you to navigate the complexity of your maintenance or repair manual, enabling you to carry out the work more easily.

WHY AN OUTBOARD?

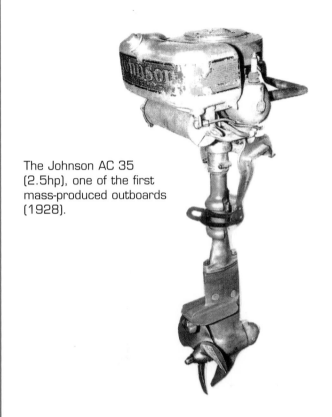

The Johnson AC 35 (2.5hp), one of the first mass-produced outboards (1928).

Nowadays, neither professional nor pleasure sailor can really safely go to sea without an engine. Outboard or inboard, each has its place.

The invention of the outboard motor revolutionised the recreational boating market. These days, few people would dispute the usefulness or efficiency of this means of propulsion.

The past

◆ 1881: Gustave Trouvé displayed the first external motor, which was electric.
◆ 1896: The first production gasoline outboard was the 'American' built by the American Motor Co of Long Island, NY. It was air-cooled, and claimed to propel a rowboat at 6–8 miles per hour.
◆ 1909: Olé Evinrude set the standard for the removable outboard with an angled leg.
◆ 1931: the 10hp Marston Seagull engine was in production in the UK: twin cylinder and capable of powering a 30ft boat.

So by the turn of the 20th century the outboard engine, as broadly defined today, was born. But it was not until after World War II that this concept gave rise to an entire industry.

Outboards today

The outboard engine has become an important piece of boating equipment thanks to its compact, stand-alone format, which combines propulsion and direction – and for its power, which today ranges from 2hp to around 300hp.

Because it is attached to the transom of a boat, it allows the maximum amount of space in the cockpit. There is no need for a rudder; direction is supplied by the flow of water created by the propeller as the engine pivots, giving excellent steering. A set of remote steering controls connected to an instrument panel allows for easy manoeuvering. Because it can be tilted, the angle can be adjusted to take into account load, speed and sea state. Since the engine can be raised, it enables a boat to be easily landed and trailed. The smaller models can be removed and carried.

Most outboards were originally two-stroke, but in order to address the pollution and fuel consumption problems of the two-stroke engine, manufacturers developed a range of four-stroke outboards.

Although the early outboard engines had a reputation for being temperamental, modern engines have now reached a very high level of mechanical reliability.

EU 2004 emissions standards labelling

EU 2005 Emissions Standard

This label indicates that the engine meets EU emissions standards (exhaust gas, noise level) proposed by the European Community Commission on 12 October, 2000. This standard came into effect on 1 January 2005.

EPA 2006 – Standard

This label indicates that the engine meets US Environmental Protection Agency emissions standards.

Two and three star labels

◆ **The two-star label** indicates that the engine meets CARB (California Air Resources Board) 2004 emissions standards.
◆ **The three-star label** that the engine meets CARB standards for 2008.

Distribution networks

There are now close to 200 different outboard engine models available on the market, which is largely dominated by American and Japanese manufacturers.

USA

The OMC group (Outboard Marine Corporation) distributes a range of 'identical' engines through its Evinrude and Johnson network.

The Brunswick group and its subsidiary, Marine Power, market Mercury and Mariner engines, which differ only in appearance from their Evinrude and Johnson counterparts.

Force, a subsidiary of the Brunswick group, markets the traditional Chrysler engines.

Japan

Along with Tohatsu, the Japanese motorcycle manufacturers Yamaha, Suzuki and Honda have provided stiff opposition to the US giants since the 1970s.

Europe

Although they had a strong presence in the market at the beginning of the century with brands such as Goiot, Alto, Archimede, Lutetia, Anzani, Watermota, the only European brands available today are Selva of Italy and the legendary British Seagull.

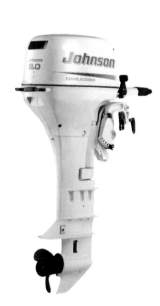

The outboard consists of three main parts:

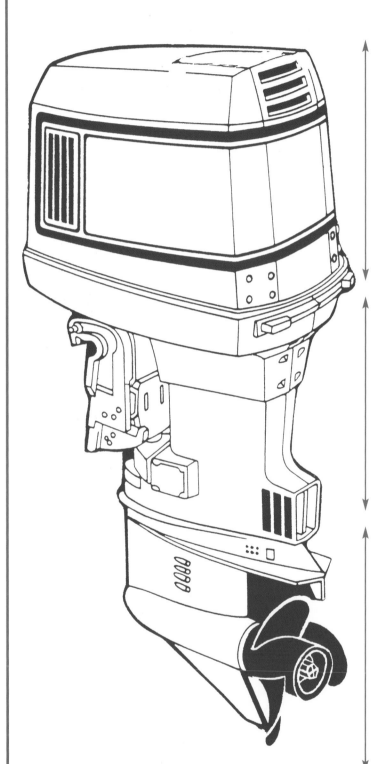

The power head

This combines the engine block and the other systems around it: ignition, fuel, lubrication, and starter.

The shaft

Also called the pivoting support, it includes the systems for trim, attachment, tilt and steering.

The lower leg

This is the propulsion part of the outboard. It primarily contains the bevel gears, the propeller and the water pump.

The environmental conditions in which the outboard engine is used demand the utmost care in the choice of construction materials. The use of specialised aluminium for the propulsion unit and high quality stainless steel for all the nuts, bolts and shafts give today's outboard resistance to the cumulative effects of marine corrosion.

Modern two- and four-stroke outboards have benefited from many technical improvements over the years, including the use of electronics, which have affected their use, performance and reliability. A result of this, however, is that maintenance and repairs are becoming increasingly complex.

HOW IT WORKS

The outboard engine is an internal combustion engine, which operates by four basic processes:

Phase 1 – Intake
Intake of air or fuel/air

Phase 2 – Compression
Compression of air or fuel/air mixture

Phase 3 – Combustion/Expansion
Rapid burning of fuel and expansion of gases

Phase 4 – Exhaust
Burnt gases are pushed out

There are three different cycles:

Four-stroke cycle: the four phases of the cycle occur in two revolutions of the crankshaft.

Two-stroke cycle: the cycle occurs in just one crankshaft revolution.

Rotary cycle: the cycle occurs three times per crankshaft revolution.

 Excessive fuel consumption and reliability problems have phased the rotary cycle engine out of the market.

Diesel outboards?

It is true that a diesel outboard can be appropriate for boats with displacement hulls, where the need for fuel economy and reliability outweighs the power/weight ratio of the engine. To meet this need, some years ago Yanmar developed 27 and 35hp engines but they were very expensive and failed to make an impact on the market.

How it works

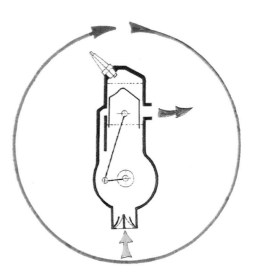

Phase 1 — Intake

Phase 2 — Compression

Phase 3 — Combustion/Expansion

Phase 4 — Exhaust

The four-stroke cycle

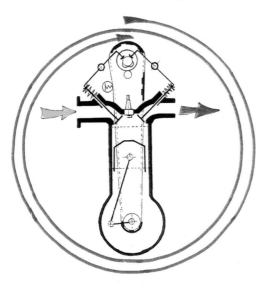

In this engine type, the four phases occur in two revolutions of the crankshaft.

The two-stroke cycle

In a two-stroke engine, all four phases occur in one crankshaft revolution.

The four-stroke cycle

In a four-stroke engine, the complete cycle requires four strokes of the piston, ie two revolutions of the crankshaft.

 The engine is lubricated under pressure by a pump that draws oil from the engine sump.

1st stroke – Intake

The piston travels down. There is an intake of gases through the open intake valve.

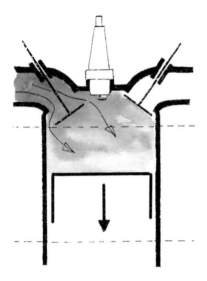

2nd stroke – Compression

The intake valve closes. The piston goes up. The mixture is compressed.

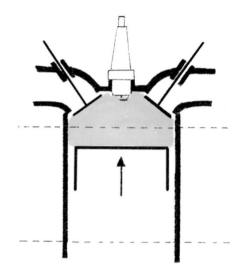

3rd stroke – Combustion

The spark plug spark ignites the gaseous mixture causing the gas to expand and push the piston. This is the power stroke.

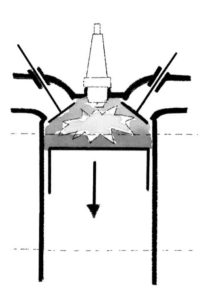

4th stroke – Exhaust

The piston moves up and expels the burnt gases out through the exhaust valve.

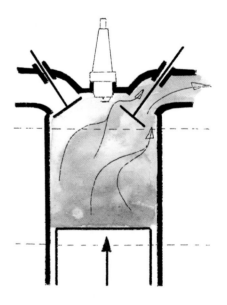

The cycle is complete; the next one can begin. It should be noted that of the four strokes in the cycle, just one produces power.

The four-stroke power head and its systems

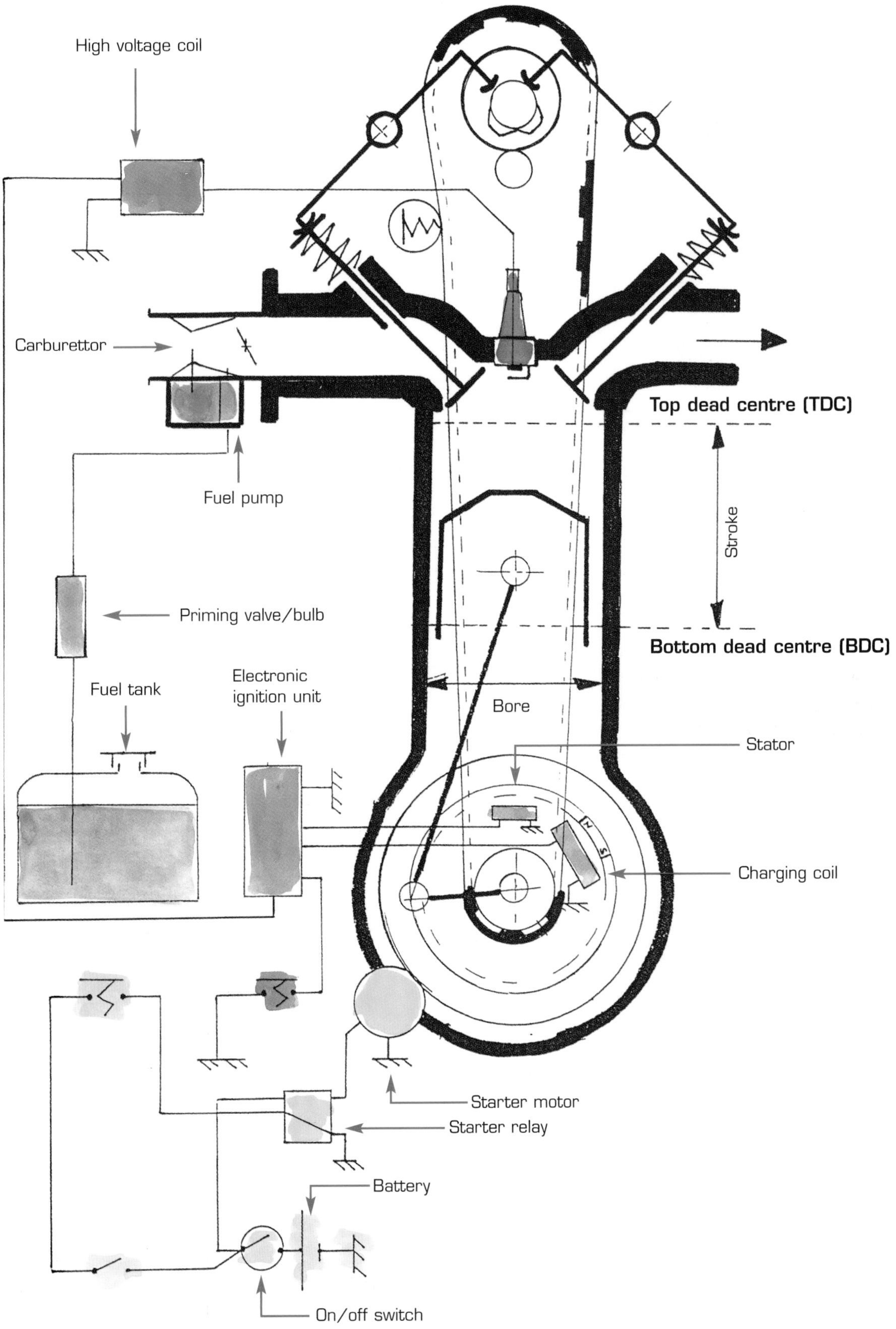

High voltage coil

Carburettor

Fuel pump

Priming valve/bulb

Fuel tank

Electronic ignition unit

Top dead centre (TDC)

Stroke

Bottom dead centre (BDC)

Bore

Stator

Charging coil

Starter motor

Starter relay

Battery

On/off switch

The two-stroke cycle

The two-stroke engine cycle can be divided into four phases, which take place in two strokes of the piston, ie one revolution.

 The engine is lubricated by the oil in the fuel or by oil injection.

Phase 1
Intake of fuel/air to the crankcase

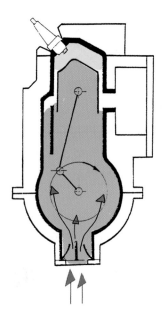

Phase 2
Compression of gas above the piston and ignition

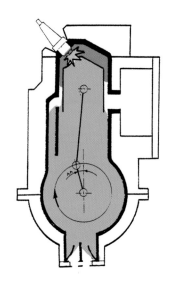

Phase 3
Power stroke

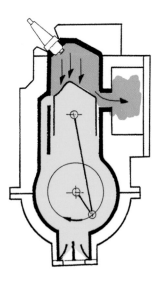

Phase 4
Exhaust of burnt gases

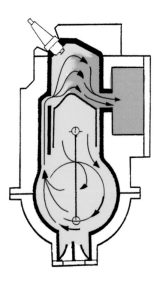

The two-stroke power head and its systems

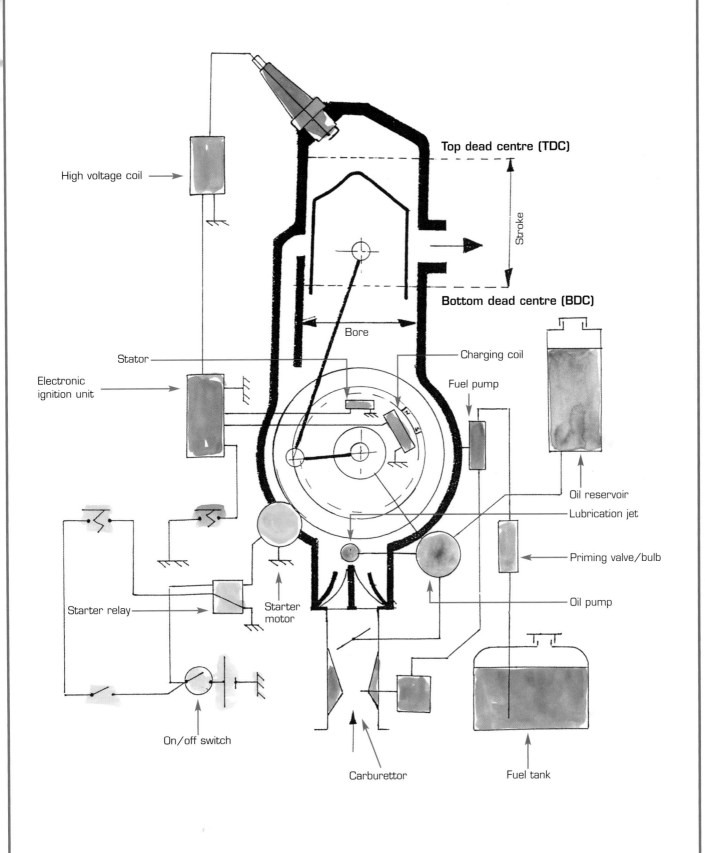

High voltage coil

Top dead centre (TDC)

Stroke

Bottom dead centre (BDC)

Stator

Charging coil

Bore

Electronic ignition unit

Fuel pump

Oil reservoir

Lubrication jet

Priming valve/bulb

Starter relay

Starter motor

Oil pump

On/off switch

Carburettor

Fuel tank

Scavenging

Being very simple mechanically, the power head of the two-stroke outboard contains far fewer moving parts than a four-stroke outboard. The intake, transfer and exhaust ports distribute the gases. However, during the process of scavenging, when exhaust gas escapes and is replaced by fresh fuel and air, some fresh and burnt gases inevitably mix, causing a loss in efficiency. In compensation, the engine benefits from one power stroke per revolution.

Cross-flow scavenging

A deflector on the piston directs the gases towards the top of the cylinder head before they escape through the exhaust port. This system, which allows smooth and economical engine performance at low revolutions, is now used only in lower power engines.
  Manufacturers prefer loop-flow scavenging.

Loop-flow scavenging

In this system the pistons are flat. The cylinders have transfer ports opposite the exhaust port. During scavenging, the exhaust pressure is used to reduce the loss of fresh gases; this increases the power output of the engine at high revs.

Lubrication

Lubrication is provided either by adding a specified amount of oil to the fuel (1 to 2% depending on the manufacturer) or by means of a separate oil reservoir that has a gravity fed variable flow pump. The oil is injected into the intake port or is distributed to various points in the sump.

Cross-flow scavenging

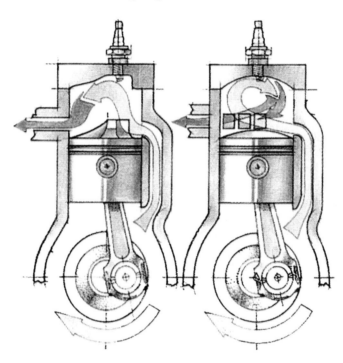

Loop-flow scavenging

Two strokes or four?

Thanks to its simplicity and consequent low cost, the two-stroke engine has been the engine of choice for the last few years. Today this is not always the case. For reasons of fuel economy as well as to satisfy emissions standards, the outboard engine has moved on.

As we have already seen, there are two options: the four-stroke engine and the two-stroke engine. Most two-stroke engine manufacturers now make four-stroke models.

With four-stroke engines of 2–300hp and fuel injected two-stroke engines of over 50hp, the days of the two-stroke engine with a carburettor are coming to an end.

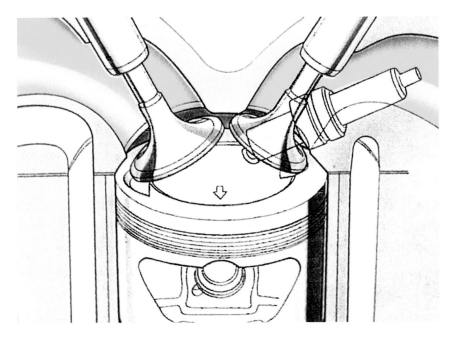

Remember, too, that the use of valves on the four-stroke engine causes an efficient separation of the intake and the exhaust cycles. This is not the case with the two-stroke engine, where exhaust ports and intake ports are open at the same time. Even though the four-stroke engine meets the new fuel consumption and emissions requirements, its complex mechanical configuration increases its retail price. It is also worth noting that its power/weight ratio is greatly inferior to that of a two-stroke engine and it requires more maintenance (oil change every 100 hours).

The new generation four-stroke engine is technically complex. Multi-valve engines have arrived; a modified bore/stroke ratio now allows 3, 4 and soon 5 valves per cylinder to fit into the combustion chamber. This, together with sequential electronic injection systems found on 40hp engines and above, gives you top performance in terms of fuel efficiency and pollution. The engines in the higher power ranges, however, come mainly from the automobile industry.

THE MAIN PARTS OF AN OUTBOARD

Four-stroke engine

The piston

The piston has a triple role. It compresses the fuel/air mix, receives the power from the combustion of gases and, in the case of the two-stroke engine, also distributes the gases during intake and exhaust. The piston head can assume various shapes depending on the type of scavenging – cross flow or loop flow.

The valves

Mounted on the engine head, sliding in a guide pushed open by the camshaft and closed by one or more springs, the valves distribute the gases in the cylinder. A machined and rectified conical seat gives the perfect air-tightness required for proper engine function.

The engine block

Its shape is dictated by the number of cylinders and their position. It contains the intake, transfer and exhaust ports.

The camshaft

On four-stroke engines, the camshaft provides the distribution of gases. It is driven in rotation by the engine and has cams that make the valves open and close. It also (usually in the case of four-stroke engines) drives the oil pump. 'Lateral' and 'head' camshafts are terms used to define the camshaft position. The lateral camshaft is located in the engine block and rocker arms and push rods are needed to operate the valves. The head camshaft, found in all four-stroke outboards, is located directly on the engine head, which eliminates the need for rocker arms and push rods.

Two-stroke engine

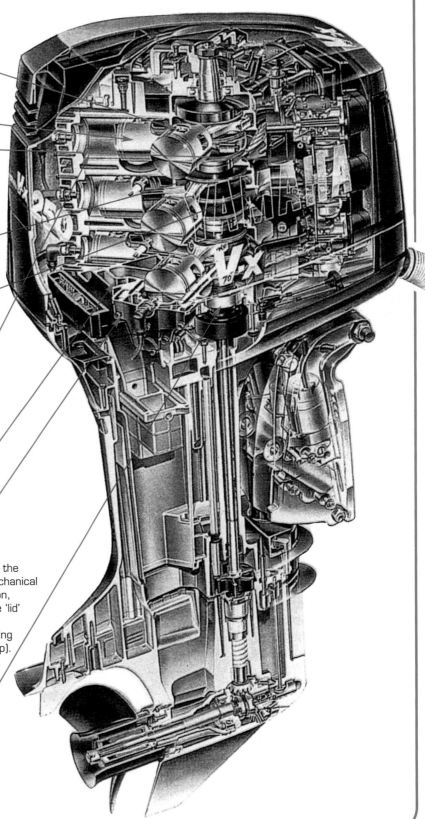

The piston pin
The piston pin connects the piston to the connecting rod.

The reed valves
The reed valves are automatic valves that open in the direction of the flow of gases during the intake stroke and close automatically when the piston travels back down during pre-compression, thus preventing recirculation of the gases.

The connecting rod
This transfers the power from the piston to the crankshaft.

The crankcase
Also called the sump, this is where the intake stroke cycle starts (the piston travels up), then pre-compression occurs (the piston moves down). It has to be perfectly air-tight. It is also the part that supports the crankshaft.

The cylinder
The cylinder is a cylindrical chamber in which the calorific energy is transformed into mechanical energy. Air-tightness is achieved in its upper end by the engine head and in its lower end by the piston rings.

The engine head
The engine head is the part that completely closes the cylinder tops. It is an area of high thermal and mechanical constraint, which occurs mostly during compression, ignition, and stroke. The two-stroke engine's simple 'lid' becomes, in a four-stroke engine, one of the most complex and intricate parts of the engine, supporting the distribution systems (camshaft, valves, oil pump).

The crankshaft
By converting the alternate linear movements of the piston(s) into rotary movement, it generates torque and transmits it to the drive shaft.

To understand how an outboard engine works, it is useful to know the meaning of terms used by their manufacturers and mechanics. When you look at engine specifications, certain terms may puzzle you: piston displacement, bore, stroke, power at propeller shaft, etc. So, here are some simple definitions.

Piston displacement

We will differentiate between a single cylinder and total cylinders.

Piston displacement – single cylinder

The volume swept by the piston between the bottom dead centre and the top dead centre, expressed in cubic centimetres, is known as piston displacement.

Piston displacement – total cylinders

This is the product of a single cylinder multiplied by the number of cylinders. When comparing different models with the same power, this can be a differentiator, even if the difference is negligible. A larger piston displacement will produce more torque at low revolutions but with slightly higher fuel consumption. A smaller piston displacement will yield a lower power/weight ratio. However, to deliver all of its power, the engine will have to run at slightly higher revolutions. Even so, when running at full throttle, the two engines' performance will be very similar. As you can see, it is not that simple!

Compression ratio

This is the ratio between the total cylinder volume when the piston is at bottom dead centre and the remaining volume when the piston is at top dead centre.

Piston displacement
Piston displacement is the volume swept by the piston(s).

Top dead centre (TDC)
The piston's uppermost position at the end of its upward stroke.

Stroke (S)
The distance travelled by the piston between the top dead centre and the bottom dead centre, it corresponds to a half revolution of the crank-shaft, ie 180°.

Bottom dead centre (BDC)
The piston's lowest position at the end of its downward stroke.

Bore (B)
The cylinder diameter.

Torque

The torque is the product of the force on the connecting rod multiplied by the crankshaft arm length. This torque is measured in Newton-metres or lb/ft.

Horsepower

Horsepower is the amount of work produced by the torque in one unit of time. Based on torque values measured on a bench, the manufacturer calculates horsepower for each type of engine with respect to the engine's revolutions per minute. The manufacturer's horsepower readings are based on the ICOMIA 28 standards measured at the propeller shaft. The current units of measurement are the watt and the kilowatt, but manuals and worksheets often refer to horsepower. 1hp = 0.746kW.

Power to weight ratio

This is the determining factor in the choice of an outboard in the small to medium power range. Those who have never walked along a pontoon with a motor in their arms should try it. They will soon appreciate the advantage of less weight.

Horsepower being equal, a boat with an outboard with a higher power/weight ratio, will plane more easily and go faster.

Specific fuel consumption

This is the weight of fuel consumed relative to the power produced in a given unit of time; or, the quantity of fuel in grams needed by the engine to produce one watt-hour of work.

The performance of outboard engines with the latest technology (electronic ignition with variable timing curve regulated for the load; ceramic fibre reinforced pistons, cylinder surface treatments, electronic injection, pre-atomised oil injection, etc) is in the 25% range, with a specific fuel consumption of between 240 and 400gr/hp/h.

Number of cylinders

We rarely find single cylinder engines these days, except for those under 5hp. It is true that a sales pitch touting multiple cylinders for the same horsepower can prevail on the basis of safety. However, for the same power, an engine with the most cylinders will run more smoothly with less vibration. All of these points are a matter of opinion.

Generally, manufacturers use the single cylinder engine for outboards under 5hp, the two-cylinder for the under 25hp models, and the tri-cylinder for over 30hp. We then go progressively into 4, 6, 8 cylinders in 'V'.

Energy loss distribution

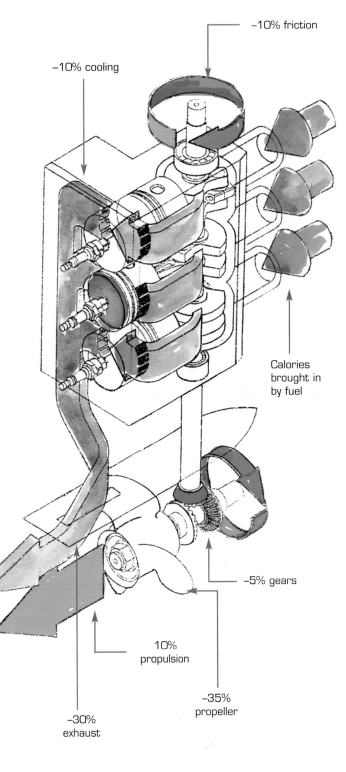

−10% friction

−10% cooling

Calories brought in by fuel

−5% gears

10% propulsion

−35% propeller

−30% exhaust

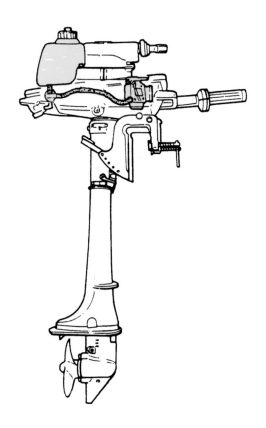

Fuel system

Safety regulations and emissions standards have considerably changed our engine's fuel system during the last decade or so. At the beginning of the 21st century, what can we expect?

Fuel circuit in low power outboards

This is the simplest system of all. The fuel, stored in an overhead tank, supplies the carburettor by way of a stopcock and flexible hose. However, the breather valve on the fuel cap must be open when starting the engine and, of course, closed during transport or storage.

Fuel circuit on outboards with a separate fuel tank

The system is a little more complicated when the fuel supply is in an auxiliary tank.

The fuel is sent to the carburettor after passing through a filter, a host of junction fittings, the manual pump or priming valve/bulb, another filter and finally the main fuel pump.

Fuel supply with built-in tank

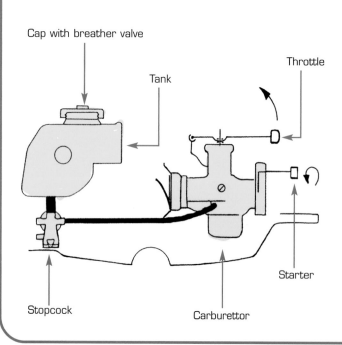

Fuel supply with separate built-in tank

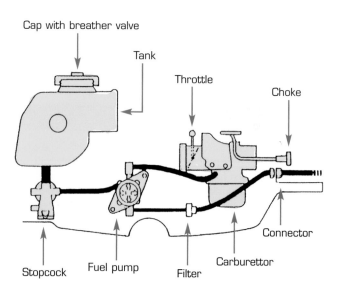

The diameter of the flexible hoses, their condition and their air-tightness at different connections all affect the smooth running of your engine. Look for leaks, and when changing flexible hoses use only those with the recommended inside diameter.

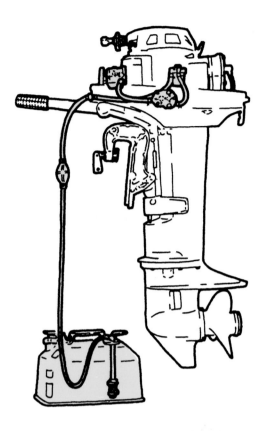

Pumps and filters

As we have seen, the fuel goes through at least two filters and two pumps before being mixed with air.

The strainer in the fuel tank is the first filter and removes the larger solid impurities. The second filter is on the engine at the fuel pump intake.

Filter maintenance

It is important to keep both filters clean.

For filters located on the tank: remove the fuel gauge then remove the filter. Clean the filter if it is clogged.

 Traces of rust on the filter mesh indicate that the tank is rusted. There is no remedy for this level of deterioration. A rusted tank is beyond repair.

Filter located on the engine

Unscrew the bowl. Check if the fuel in the bowl contains water. Clean the filter by soaking in petrol or, if deteriorated, replace it.

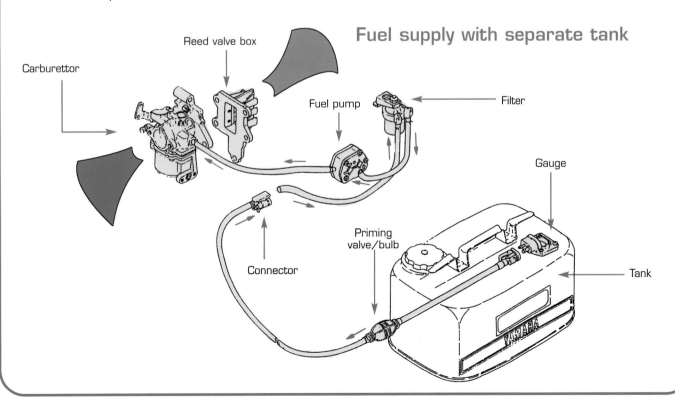

Carburettor

Reed valve box

Fuel pump

Filter

Fuel supply with separate tank

Gauge

Priming valve/bulb

Connector

Tank

The connectors

Although each manufacturer uses different connectors, their basic principle is identical. Whenever there are air or fuel leaks, check the condition of the snap connector and ball system. If they are in poor condition, the whole connector must be replaced.

The priming valve/bulb

Its role is vital at engine start-up but also while the engine is running. When first starting, a few squeezes of the bulb should be sufficient to prime the fuel system.

> ✴ **The bulb should feel firm and stay firm. A lack of pressure retention indicates a fuel leak. Check the bulb valves and the various connectors. Be aware that a leaking carburettor needle valve causes the same symptoms.**

Fuel pump at 'intake' phase Piston upstroke

Pc = pressure in the crankcase
Pa = atmospheric pressure
Pp = pump pressure

The piston rises. Intake of air/fuel to the crankcase

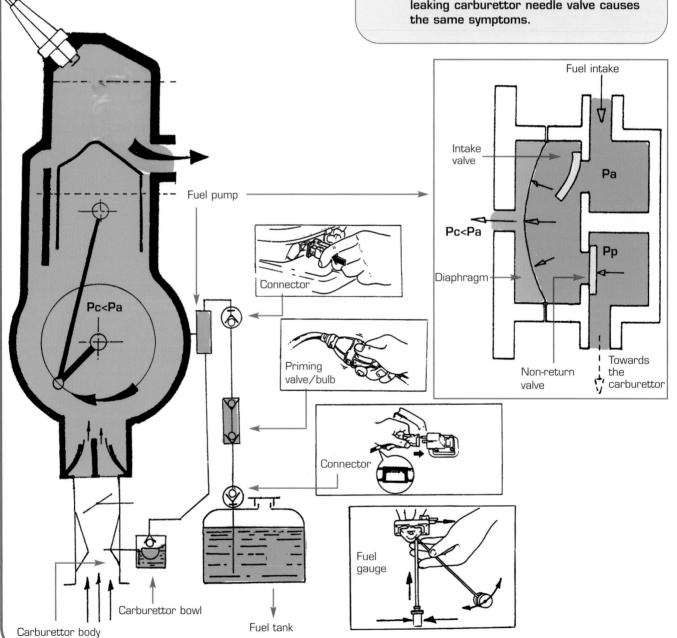

The fuel pump

On outboards, a diaphragm type of fuel pump is used. Attached to the engine or integrated into the carburettor, it is made to work by pressure variations in the crankcase. When the piston travels upwards the pressure is negative. This is the intake phase. When the piston travels downwards, the pressure is positive. This is the exhaust phase.

◆ **Maintenance**
   The fuel pump is fragile and expensive. Be very careful when taking it apart.

◆ **Inspection**
   Before removing the fuel pump, check for possible external leaks.

When disassembling it, check the diaphragms and the condition of the valves. There is damage to the diaphragm if the base of the spark plug tends to remain damp or when the fuel supply to the carburettor is abnormal. Replace the diaphragm if it is severely deformed.

Fuel pump in 'exhaust' phase: piston downstroke

Pc = crankcase pressure
Pa = atmospheric pressure
Pp = pump pressure

The piston travels down. Pre-compression in the crankcase

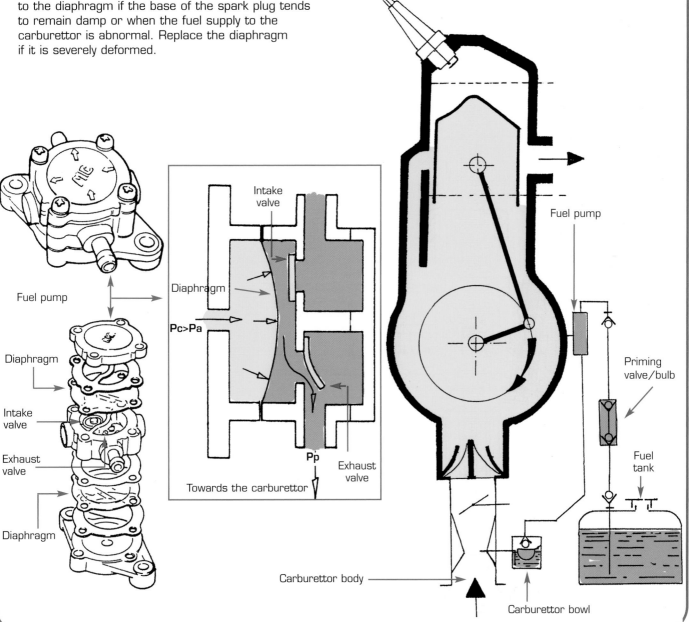

Fuel pump

Diaphragm

Intake valve

Exhaust valve

Diaphragm

Intake valve

Diaphragm

$Pc>Pa$

Pp

Towards the carburettor

Exhaust valve

Fuel pump

Priming valve/bulb

Fuel tank

Carburettor body

Carburettor bowl

The carburettor

The carburettor is the means of supplying fuel to engines running on a fuel/air mixture.

The amount of fuel provided is regulated by the throttle and, for cold starts, a manual or electric choke.

A carburettor consists of two main elements: the bowl and the body. The former, which is fed by the fuel pump under very low pressure, contains the fuel, which is kept at a constant level by the float/needle assembly.

Adjusting the mix

The carburettor regulates the fuel/air ratio (1g/15g) and also regulates the quantity of the mix.

The dosage is delivered by the jets and different circuits in accordance with the engine revolutions.

The throttle slide carburettor

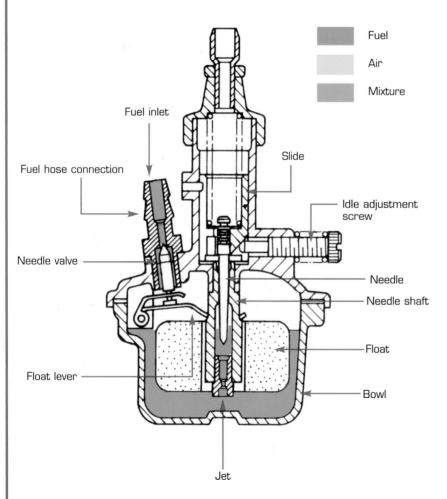

Fuel inlet

Fuel hose connection

Needle valve

Float lever

Jet

Slide

Idle adjustment screw

Needle

Needle shaft

Float

Bowl

- Fuel
- Air
- Mixture

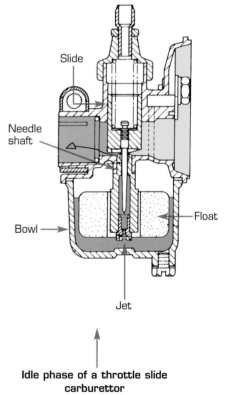

Slide

Needle shaft

Bowl

Float

Jet

Idle phase of a throttle slide carburettor
The slide rests against the stop screw. The fuel drains into the bowl via the needle valve. The fuel level in the bowl is determined by the float level adjustment. The flow of fuel at idle is controlled by the position of the stop screw and the needle height.

So, what we have is: one jet on the idle circuit, accessible from the exterior of the carburettor, which governs the richness of the idle mix, and a main jet with an emulsifying tube that handles the higher revolutions. The idle circuit works together with the progression circuit (which consists of small holes upstream of the throttle valve) to manage the transition from idle phase to what I will call normal running speed.

Setting the richness of the mix at idle is done by adjusting the idle screw, which, like a tap, regulates the airflow in the circuit. Unscrew it and you increase the quantity of air, ie you make the mix leaner. Screw it in and the reverse happens. However, a medium mix is easy to find. Slowly turn the idle adjustment screw all the way in, then unscrew it one and one half turns. On multi-cylinder engines, manufacturers generally use one carburettor per cylinder or a double-barrelled carburettor for two cylinders. These carburettors are controlled by a butterfly valve for each sector. Synchronisation is achieved by modifying the length of the control rods.

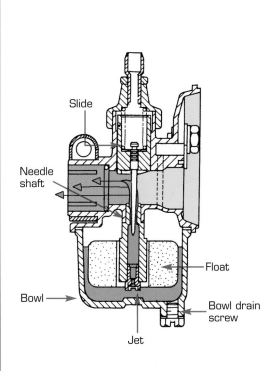

Fully open phase of a piston carburettor

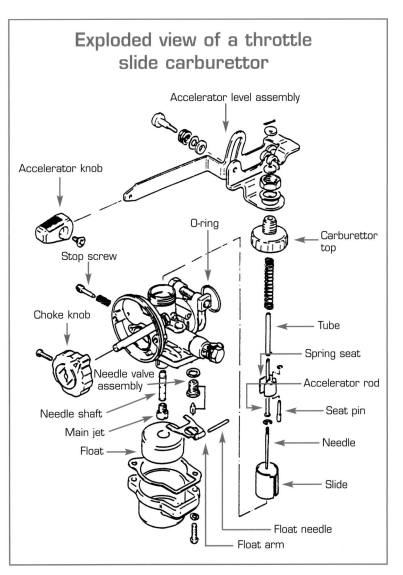

Exploded view of a throttle slide carburettor

Cold starting

For cold starting, the fuel/air mix must be made much richer.

While some outboards actually have pumps that inject fuel behind the throttle valve, the majority of outboards only have a butterfly choke valve placed upstream of the carburettor, which completely shuts off the air supply. On some models, a valve on the butterfly choke valve opens progressively after the engine starts to reduce the mix richness as the revs increase.

The throttle valve carburettor

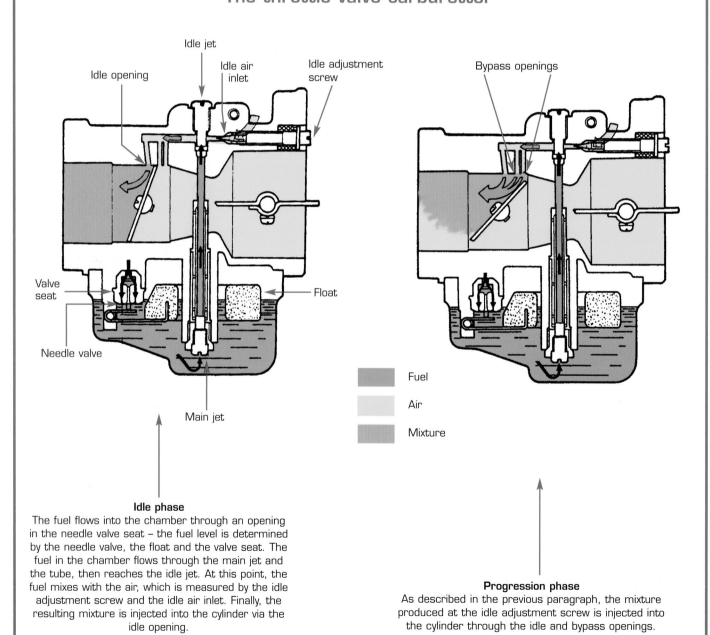

Idle phase
The fuel flows into the chamber through an opening in the needle valve seat – the fuel level is determined by the needle valve, the float and the valve seat. The fuel in the chamber flows through the main jet and the tube, then reaches the idle jet. At this point, the fuel mixes with the air, which is measured by the idle adjustment screw and the idle air inlet. Finally, the resulting mixture is injected into the cylinder via the idle opening.

Progression phase
As described in the previous paragraph, the mixture produced at the idle adjustment screw is injected into the cylinder through the idle and bypass openings.

Carburettor maintenance

Even though removing the carburettor is not difficult, it is still wise to take precautions and to work carefully and methodically when dismantling it.

Basic maintenance consists of removing the bowl, the needle valve and the various other jets to clean them with petrol. Various kinds of deposits – water, oil, rust, etc can cause serious problems.

 It is essential to have a source of compressed air to blow the jets and various circuits clean.

Main air inlet port

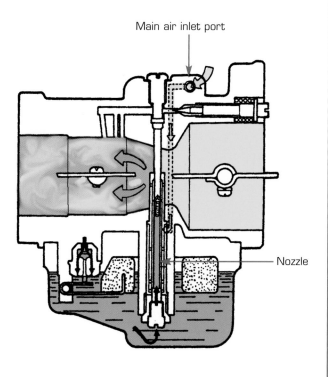

Nozzle

Full throttle phase
The main jet allows fuel to enter the carburettor chamber. Inside the chamber, the fuel is mixed with air, which has been measured by the main air inlet port. The resulting mixture is then injected into the cylinder via the main chamber.

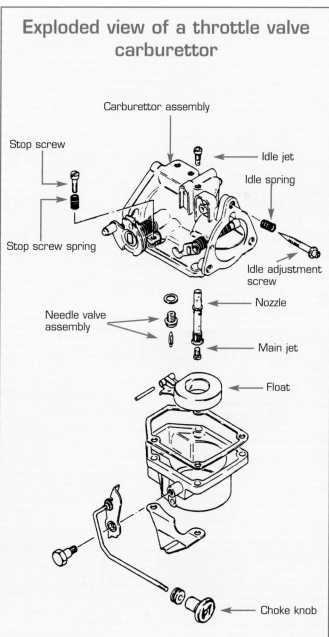

Exploded view of a throttle valve carburettor

Carburettor assembly

Stop screw

Idle jet

Idle spring

Stop screw spring

Idle adjustment screw

Nozzle

Needle valve assembly

Main jet

Float

Choke knob

Fuel injection

Many manufacturers have already adopted fuel injection to improve fuel economy and compliance with pollution standards. Fuel injection offers several advantages, among them:

◆ Lower fuel consumption
◆ More efficient starting
◆ Reduced pollution emissions
◆ Increased power
◆ Better performance

When fuel is injected directly into the combustion chamber, there is no scavenging waste and fuel consumption is therefore reduced by 30%.

Injector flow is determined by a constant analysis of engine parameters, such as the revolutions, the air intake volume, the position of the throttle control, the engine temperature, the external air temperature, etc. The quantity of fuel injected corresponds exactly to the needs of the engine. Pollution is minimal because the quality of the fuel/air mixture ensures good combustion.

However, maintenance of this carburation system requires specialised tools and highly qualified technicians.

Tohatsu TLDI system

Two-stroke with low pressure direct injection

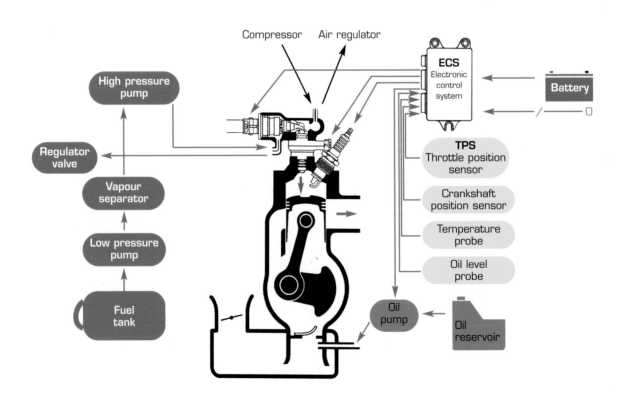

Diagram of Suzuki multipoint sequential electronic fuel injection (DF250)

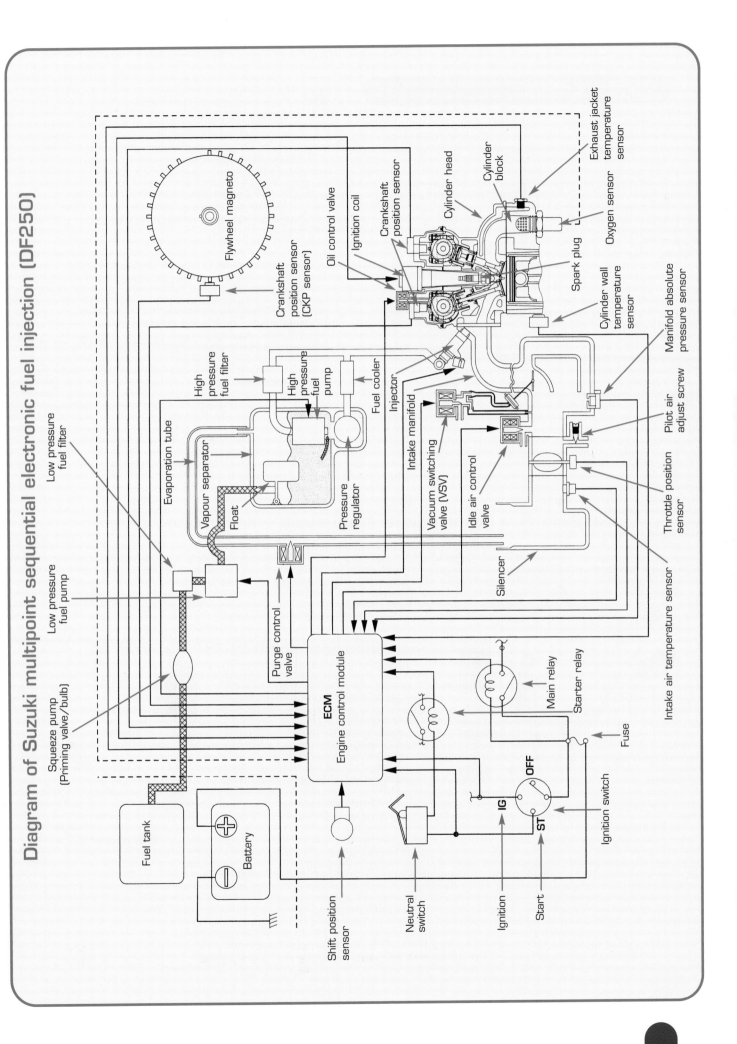

Ignition system

Thanks to advances in electronics, the ignition system is certainly one of the engine components that has most benefited from improvements in the past few years.

The ignition system sparks combustion of the air and fuel mixture compressed in the combustion chamber. To do this, each system element has a highly specialised role, ie:

◆ The power element. Composed of the flywheel magneto bolted directly onto the crankshaft head, the coil and the electronic module.
◆ High voltage coils and spark plugs produce the high voltage energy needed to create a 40,000 volt spark.
◆ The ignition triggering element with its impulse generators or sensors, determines the precise moment the spark has to ignite the mixture.

 Each of the elements of the ignition system should always be in perfect condition, especially those that are prone to wear, foremost among these being the spark plugs and, for older engines, the breakers – better known as the points. These serve to alternate the current coming from the coil to the spark plug. Today, they have been replaced by the electronic control unit, which has simplified the system so that maintenance is limited to replacing the spark plugs and keeping all of the contacts perfectly clean. Maintaining a perfect 40,000 volt spark will ensure that you have good starting as well as good combustion and reliable function at every rpm. The chronic weaknesses of old outboard engines have now largely been resolved.

Diagram of a second-generation electronic ignition system

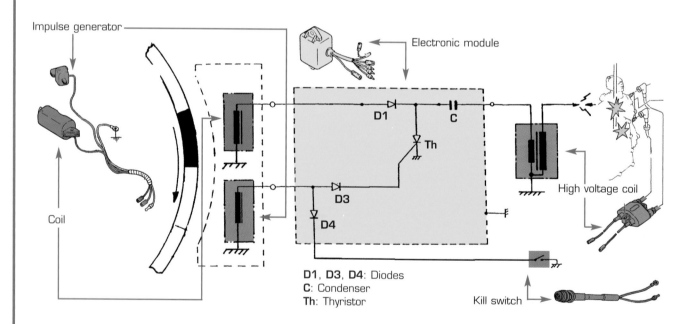

Impulse generator

Electronic module

D1

C

Th

High voltage coil

Coil

D3

D4

D1, **D3**, **D4**: Diodes
C: Condenser
Th: Thyristor

Kill switch

Classic ignition system – second generation

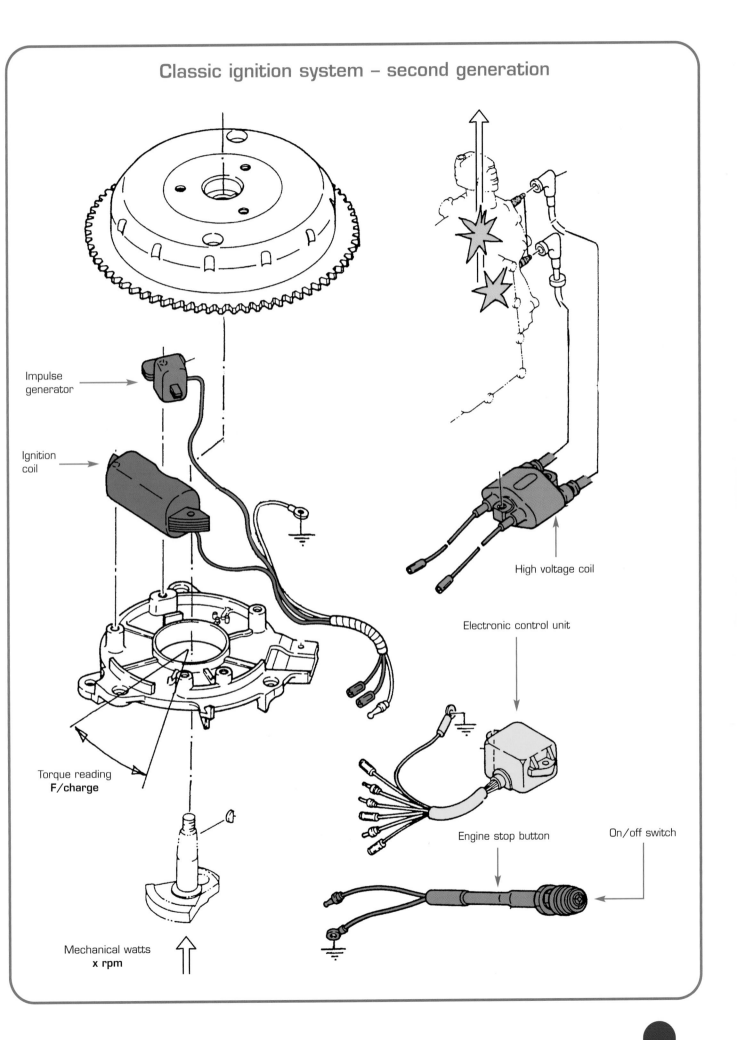

Impulse generator

Ignition coil

Torque reading
F/charge

Mechanical watts
x rpm

High voltage coil

Electronic control unit

Engine stop button

On/off switch

Outboard Wiring Colours

Old Standards	New colour code	What are they for?
Black	Black	All ground
Black	Brown electrode	Mercathode reference
Black	Orange	Mercathode anode electrode
Blue	Blue/white	Trim up switch
Brown	Grey	Tachometer signal
Grey	Yellow/black	Choke
Green	Green/white	Trim down switch
Orange	Black/yellow	Stop
Pink	Pink	Fuel gauge sensor
Purple	Purple/white	Trim trailer switch
Red	Red	Unfused wires from battery
Red	Red/purple	Fused wires from battery
Red	Red/purple	+12v fused wire to trim panel
White	Purple	+12v ignition switch
Green/white	Beige	Temperature alarm return
White/purple	Brown/white	Trim sensor to trim sender
Yellow	Yellow	Starter solenoid
Yellow	Red/yellow	Starter safety switch, when not in neutral

Capacitor discharge electronic ignition

Also known as CDI (capacitor discharge ignition), this ignition system is made up of charge elements, accumulation condensers, impulse generators and thyristors that serve as breakers.

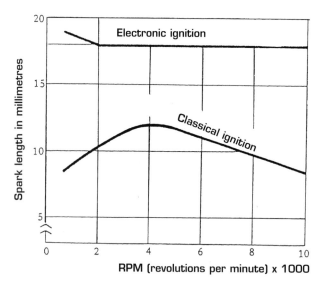

The electronic ignition spark is more powerful and remains constant regardless of the engine rpm. This is not the case with the conventional ignition spark, which reaches its maximum at 4000rpm and then decreases.

How it works

When the flywheel turns, the magnetos induce an alternating current in the ignition coil. This is inversed (that is, with the help of diode **D1** (page 34), the negative alternations are suppressed) and charges the condenser **C**. At the point of ignition, the impulse generator, or captor, gives positive and negative impulses. The negatives are suppressed in the same way as previously described (diode **D3**). The positives command the thyristor (**Th**). It becomes the conductor. The condenser discharges to earth. This abrupt variation is transmitted by the condenser to the primary coil winding, which generates a very high induction voltage in the secondary coil winding. The engine is switched off by earthing the impulses through the diode **D4** and the stop button. The thyristor is no longer receiving orders; there is no more ignition.

 The voltage produced by the ignition coil depends on the coil's elevator ratio and the current's rate of variation, ie the engine rpm.

It is difficult for an ignition coil to supply a charge current that is satisfactory at both high and low rpm. For this reason, on some models manufacturers use two ignition coils – one for low revs and one for high revs – to achieve a constant wave of high voltage at any level of rpm. For the same reason, in most cases each cylinder has its own high voltage coil.

With these ignition systems breakdowns are reduced, but it is impossible to make repairs as all elements of the electronic module are imbedded in resin and enclosed in a sealed box.

If the electronic block fails it must be replaced. However, after replacing it and before re-starting the engine, the cause of the breakdown must be identified to avoid destroying the electronic module again.

Advantages of this type of ignition

◆ Constant and maximum high voltage, whatever the revs.
◆ Very fast rise in voltage, which has the effect of providing greater protection against spark plug fouling and insulator problems.
◆ Extended spark plug life (no fouling).
◆ Since CDI electronic ignition doesn't have breaker points that are constantly subject to wear and tear, maladjustment (and therefore maintenance) is completely eliminated.
◆ The improvement in ignition power (40,000 instead of 15,000 volts) and the absence of breaker points means that the CDI ignition system is not affected by even extreme humidity (provided, however, that the connections are well insulated).

The spark plug

The plug is the only part of the ignition system that has still retained its old vices – fouling and getting out of adjustment. Because of this, it needs constant maintenance. Plugs should be removed and checked after about 50 hours of operation.

Spark plug structure

The spark plug consists of an electrode-tipped rod embedded in a ceramic insulator, which is in turn inserted into a threaded metal base with a ground electrode.

The upper part of the base is forged in the shape of a hexagonal nut. The part of the base above the threaded section may be conical or flat. In the latter case, it contains a compression-sealing washer.

Which spark plug do I use?

Outboards are differentiated from one another by their horsepower, their ignition type, their cooling system, the shape of the combustion chamber, their fuel system, etc. It is therefore impossible to use the same spark plugs on all types of engines. Furthermore, spark plugs have to be appropriate for each type of use.

The main criteria are :

◆ Dimensions (thread diameter, base length, reach, type of seat)
◆ Thermal rating, ie, the plug's capacity to dissipate heat accumulated in the central core with each explosion.
◆ The way they are made, the type of electrode, integrated resistance, etc.

Anatomy of a spark plug (NGK doc)

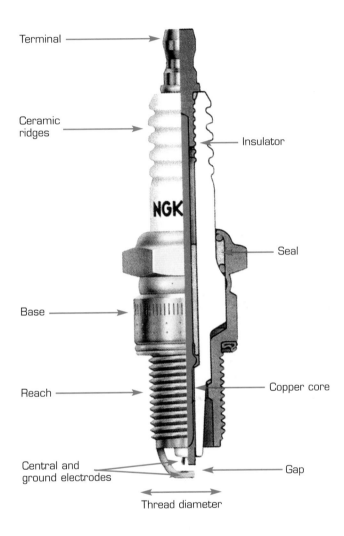

Terminal

Ceramic ridges

Insulator

Base

Seal

Reach

Copper core

Central and ground electrodes

Gap

Thread diameter

Different thread reach (NGK doc)

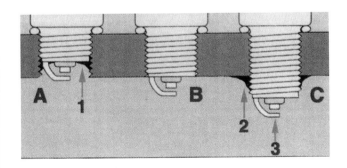

A Reach too short
1 Risk of carbon deposits forming on the threads
B Correct reach
C Reach too long
2 Risk of carbon deposits forming on protruding part of thread
3 Ground electrode at risk of overheating or hitting the piston

Thermal rating

It is essential that the spark plug rapidly reaches its self-cleaning temperature, which is about 400° to 500°C. When it does, residues such as carbon are burnt off; the insulator cone remains clean. To avoid self-ignition, it is vital that the temperature doesn't exceed 850°C at full rpm. A rise in electrode temperature can also cause a piston to be pierced. Practically speaking, we can say that small, low-powered engines that have low combustion temperatures require 'hot' spark plugs with a low thermal rating. Conversely, larger, high powered engines, which have high combustion temperatures, require 'cool' spark plugs with a high thermal rating.

Spark plug appearance

Careful examination of the electrodes and the insulator after a period of use can tell us how the engine is functioning.

Normally, the insulator cone should be light yellow to brownish grey in colour and the electrodes should be light grey.

Overheating will result in a whitish coloured insulator. This may be caused by the timing being too fast, a fuel/air mix that is too lean, or a spark plug with too low a thermal rating.

Black carbon deposits on the insulator cone cause the engine to misfire, resulting in poor engine performance.

Oily deposits on the cone are a sign that the spark plug is operating at too cool a temperature. First check:

◆ that the spark plug thermal rating is correct for your engine type and operating conditions;
◆ the oil pump adjustment or percentage of oil in the fuel mix;
◆ that the oil quality is appropriate for two-stroke engines.

 On a four-stroke engine, an oily deposit can be caused by a worn head gasket. It can also be caused by worn out rings or oil seals on the valves.

Difficult starts and misfiring can be caused by using the engine for too long at low revs or by a fuel mixture that is too rich.

Important points

◆ Use spark plugs with the correct thermal rating.
◆ Check the electrode gap.
◆ Follow the installation instructions.
◆ Replace the spark plugs every 200 hours or at least at the beginning of each season after winter storage.

For better performance check your spark plug

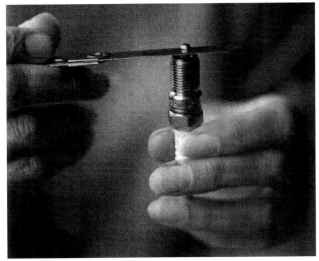

Electrode gap adjustment is done with a feeler gauge.

Carbonised spark plug (Champion).

Overheated spark plug (Champion).

Overheated spark plug with a shiny insulator cone (Champion).

Melted or puffy deposits (Champion).

Broken insulator (Champion).

Cooling water circulation on a low power outboard

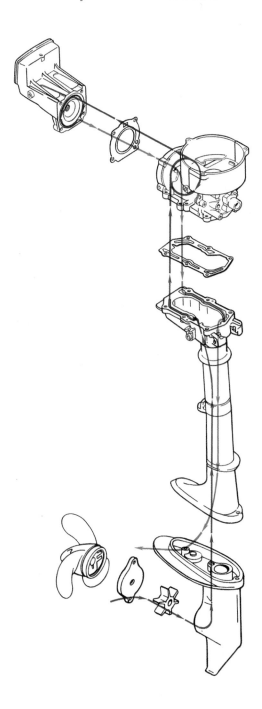

The cooling system is reduced to its simplest form. Here, the propeller shaft drives the water pump.

Cooling system

The outboard engine's cooling system dissipates the thermal energy caused by fuel combustion, and to a lesser degree friction caused by the many moving parts. Good cooling is therefore essential to the performance and preservation of the engine.

Not long ago, low-powered engines were air cooled. Despite its great simplicity and efficiency, this system is no longer used. Manufacturers now prefer direct water cooling, where water is pumped from outside the boat and is circulated around the cylinders and through the cylinder head before being expelled back out mixed with exhaust gases. A simple thermostatic valve (thermostat) regulates the engine temperature.

The thermostat opens and closes in relation to the water temperature. When the engine is cold, the thermostat is closed; no water circulates in the cylinder head, which reduces the time it takes for the engine to get up to temperature. When the engine reaches the correct temperature, the thermostat opens, the water circulates, the engine is cooled. The temperature goes down and the cycle starts again.

The water pump

A water pump, driven by the drive shaft, circulates water through the cooling system. This 'volume' type of pump consists of a rubber impeller, attached to the drive shaft by a key, spinning inside the pump casing which is offset in relation to the shaft. This type of self-priming pump is not affected by small solid particles sucked in with the water. Its only shortcoming is that it can't function without water. The rubber blades heat up and deteriorate in seconds. It is therefore wise to take certain precautions before installing and starting the outboard, most importantly:

◆ ensure that the cooling water inlets are not clogged;
◆ check the tell-tale flow.

Checking system function

A tell-tale jet of water located at the rear of the engine is a good indication that the cooling system is functioning properly. But be aware that the tell-tale jet flow does not always give a definitive indication of water circulation in the engine block, where one or more passages could be obstructed. For the same reason, the tell-tale jet temperature isn't always a reliable indicator. It is better to check the water temperature at the outlet situated in the middle of the lower leg.

Cooling water circulation in an outboard engine

High performance engines have a thermostat and a bypass valve to regulate the engine temperature. The thermostat ensures a rapid rise in temperature at low revolutions. The bypass valve activates the water flow at high revolutions to ensure greater cooling.

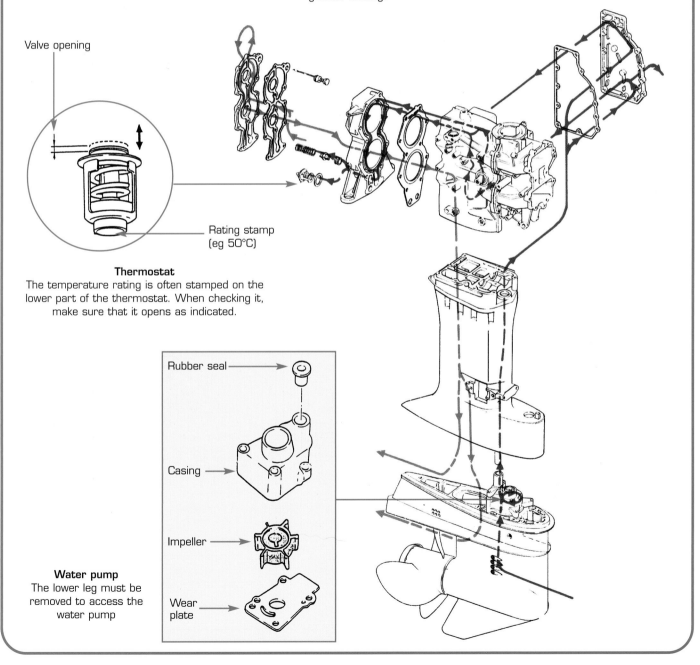

Valve opening

Rating stamp (eg 50°C)

Thermostat
The temperature rating is often stamped on the lower part of the thermostat. When checking it, make sure that it opens as indicated.

Rubber seal

Casing

Impeller

Wear plate

Water pump
The lower leg must be removed to access the water pump

Exploded view of an inversion gearbox system

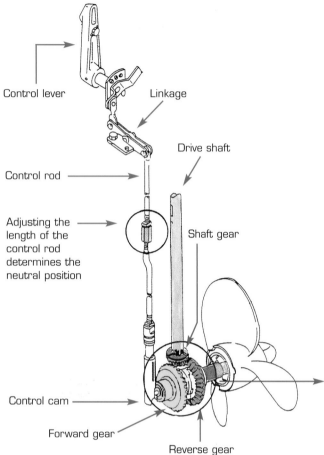

Control lever

Linkage

Control rod

Drive shaft

Adjusting the length of the control rod determines the neutral position

Shaft gear

Control cam

Forward gear

Reverse gear

This is the part of the outboard called the lower leg, which comprises several systems:

- The reduction and inversion gears
- A part of the cooling system with the water pump
- The actual drive system with the propeller and, in some rare instances, a jet

Anatomy of the lower leg

The lower leg consists of a streamlined gearbox housing, a skeg, and an anti-ventilation plate, also called an anti-cavitation plate – the height of which is crucial when fitting the outboard onto the transom.

It also contains the cooling system water inlet, the drain and fill holes for the gearbox, and the exhaust outlet.

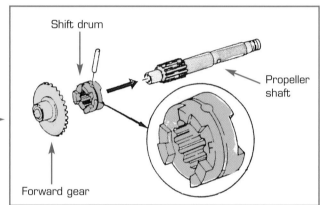

Shift drum

Propeller shaft

Forward gear

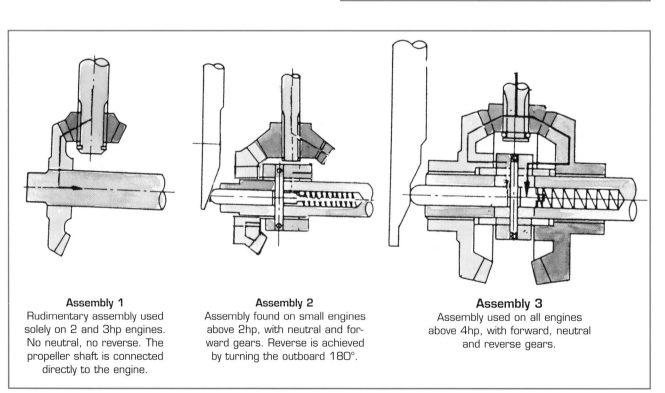

Assembly 1
Rudimentary assembly used solely on 2 and 3hp engines. No neutral, no reverse. The propeller shaft is connected directly to the engine.

Assembly 2
Assembly found on small engines above 2hp, with neutral and forward gears. Reverse is achieved by turning the outboard 180°.

Assembly 3
Assembly used on all engines above 4hp, with forward, neutral and reverse gears.

The gearbox

Located between the drive shaft and the propeller shaft, it has two functions:

- Reduction
- Inversion

Reduction

Given how fast the engine revs, a reduction gear is needed to maintain acceptable propeller performance.

To do this, manufacturers use reduction ratios that reduce the propeller speed by half. Some engines use greater reduction ratios, close to one third, to be able to use larger diameter propellers.

Inversion

This is achieved with a set of gears housed in an oil-filled gearbox located in the lower leg. Oil seals ensure the impermeability of the box.

The gear assembly consists of a shaft gear driven by the drive shaft, two counter shaft gears on the propeller shaft, and a shift drum connected to the shaft. The gears are shifted when the shift drum is moved either by a set of control rods, linkage and fork, or a cam and push rod.

 Under this noisy but effective rudimentary system, the gears must only be shifted when the engine is at idle.

Neutral (Suzuki)

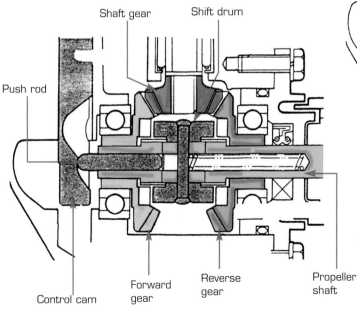

Shaft gear — Shift drum
Push rod
Control cam — Forward gear — Reverse gear — Propeller shaft

Forward gear (Suzuki)

The cam moves the shift drum laterally towards the forward gear. **A** is coupled to **B**. The gear then drives the propeller shaft.

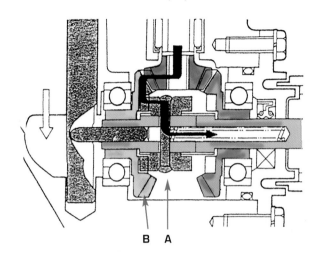

B   A

Reverse gear (Suzuki)

The cam moves the shift drum over to the reverse gear. **A** is coupled to **C**. The reverse gear then drives the propeller shaft.

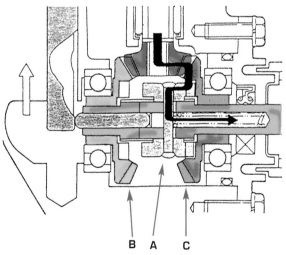

B   A   C

Propeller-to-shaft assembly

On small outboards, the propeller is locked to the shaft by a shear pin. If one of the propeller blades strikes something, the pin shears off. It is wise to have one or two spares.

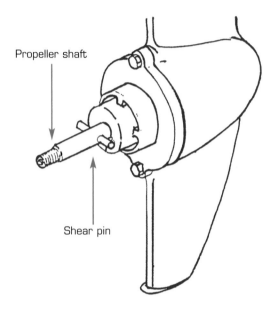

Propeller shaft

Shear pin

Shaft	Length
Short	380mm
Long	508mm
Extra long	635mm
XX long	762mm

Propeller attachment

The propeller can be attached in two ways:

◆ On low-powered outboards, a shear pin through the shaft makes the connection. The propeller is held on the shaft by a castle nut, locked in place by a cotter pin.

◆ When motoring, if the propeller strikes something, the shear pin snaps and the propeller is no longer driven, saving the gearbox. Spare shear pins and cotter pins are supplied with the outboard. They are usually found under the engine cover.

◆ To absorb accidental shock, on higher powered outboards, a neoprene bush compressed between the propeller and the hub replaces the shear pin. If the propeller strikes something, it spins around the neoprene bush, which absorbs the blow and reduces damage to the propeller and gearbox.

A Nyloc nut or a castle nut and cotter pin hold the propeller in place.

 When simple water resistance causes continuous slippage, the propeller must be replaced.

Short, long and extra-long shafts

To adapt the drive system to all boat types (from the nearly flat-bottomed Zodiac to the V hull runabout to the sailboat), outboard manufacturers offer different shaft lengths.

Propeller-to-shaft assembly

When the exhaust is through the propeller hub, the propeller is attached to the propeller shaft with a nut and cotter pin, or a screw and lock washer. When the propeller strikes something, the shock absorber, which consists of a neoprene bush inside the propeller, slides.

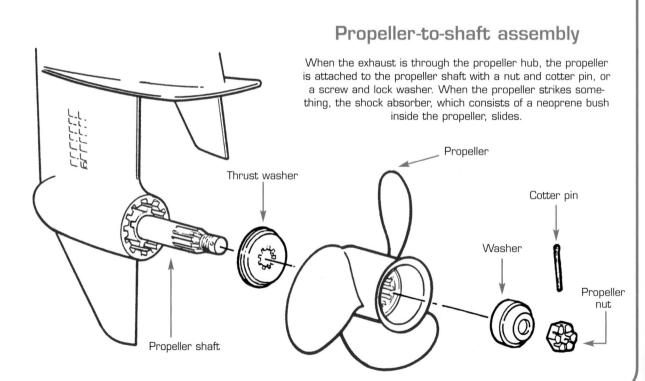

Thrust washer

Propeller shaft

Propeller

Cotter pin

Washer

Propeller nut

The propeller

It is the propeller that determines an engine's power because it converts the propeller shaft torque into propulsion or thrust. To obtain optimum thrust, the propeller's characteristics must be suitable for the engine, the hull and also the speed at which it is to be used.

A propeller consists of blades evenly positioned around a hub.

Various terms describe and characterise specific aspects of the propeller: diameter, pitch, slippage, cavitation, etc. All of these terms crop up in advertising brochures and manuals, and even in conversation with your service engineer. But what do they mean?

Some simple definitions

◆ **Diameter** This is the circle described by the blade tip when the propeller turns. For a given propeller type, the diameter will be greater for a slow boat than for a fast one.
◆ **Pitch** The pitch is the theoretical distance travelled by the propeller in one turn. To guarantee maximum engine efficiency, the pitch has to be suited to the boat's effective speed and use, eg racing, water-skiing, fishing, etc.

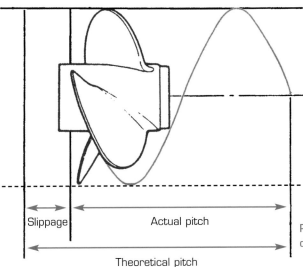

Slippage | Actual pitch

Theoretical pitch

Propeller slippage diagram

 When the propeller turns in the water, the liquid mass slips away because it does not have the same resistance as a solid body to the blades' push. The propeller throws back a volume of water, which in turn produces the thrust. The difference between the pitch and the actual distance travelled in one turn of the propeller is the slippage.

Slippage can reach 30% on displacement hulls, 25% on semi-displacement hulls, 15% for planing hulls and less than 10% on racing boats.

Main characteristics of the propeller

A propeller is described in two dimensions, expressed in inches and fractions of inches.

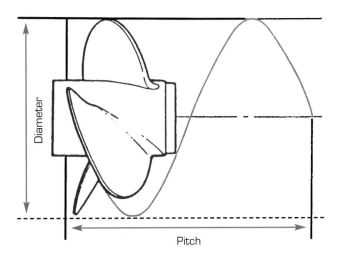

Diameter

Pitch

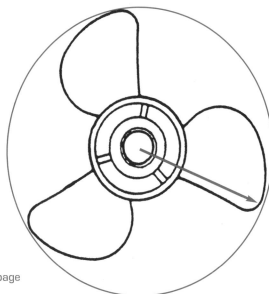

Propeller diameter
This can sometimes be difficult to read. If this is the case, measure the distance between the centre of the hub and the tip of a blade and multiply by two.

Parts of the propeller

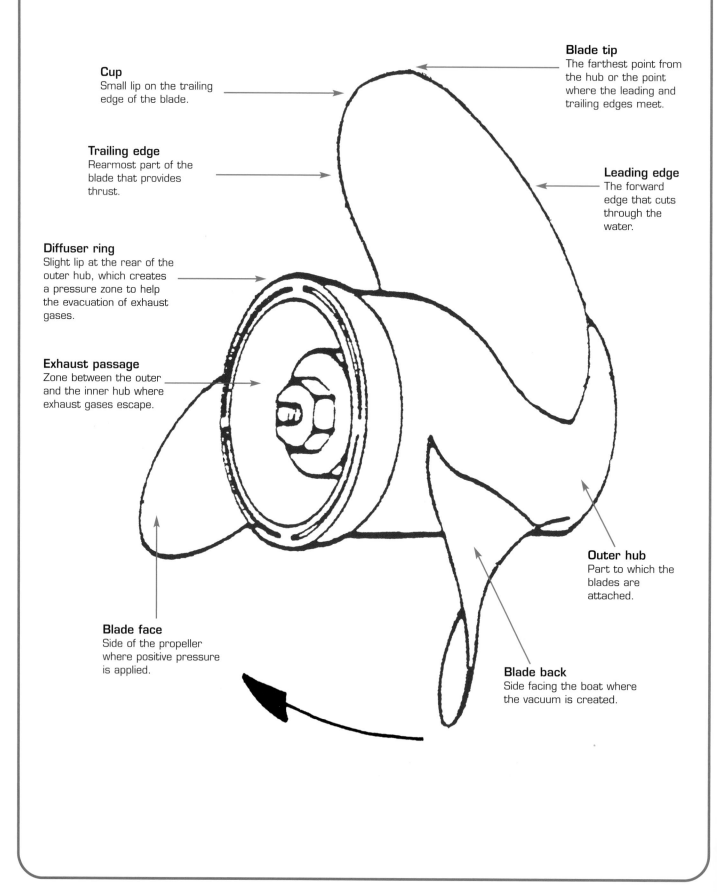

Cup
Small lip on the trailing edge of the blade.

Blade tip
The farthest point from the hub or the point where the leading and trailing edges meet.

Trailing edge
Rearmost part of the blade that provides thrust.

Leading edge
The forward edge that cuts through the water.

Diffuser ring
Slight lip at the rear of the outer hub, which creates a pressure zone to help the evacuation of exhaust gases.

Exhaust passage
Zone between the outer and the inner hub where exhaust gases escape.

Outer hub
Part to which the blades are attached.

Blade face
Side of the propeller where positive pressure is applied.

Blade back
Side facing the boat where the vacuum is created.

Choosing a propeller

Most low-powered outboards are equipped with a propeller that has a diameter and pitch that the manufacturers consider suitable for general use. This propeller can reach full power on an 'average' boat with an 'average' load. In other words, each boat, depending on its weight, length, power and use must be considered a special case, requiring a propeller with a diameter and pitch designed for it.

The choice of pitch for a given application will affect the life span of your engine. Increasing or decreasing the pitch affects the engine revs. To determine the correct pitch for your propeller and taking into account your engine's maximum horsepower, you must test it in the water. The propeller with the correct pitch will absorb the total engine power at maximum revs. If the pitch is too small, the engine will over-rev; if it is too large, the engine will not reach its maximum speed. As a general rule, an increase or decrease in propeller pitch corresponds to a difference in acceleration of around 300 revs.

Number of blades

The number of blades determines the propeller's performance and, to a lesser degree, its level of vibration. Let's say for simplicity's sake that a greater number of blades allows a smaller propeller diameter but this also reduces performance. All manufacturers have adopted the three-blade propeller as a standard. Each propeller blade has a leading edge that cuts through the water in forward motion and a trailing edge opposite it.

Propeller type

Depending on your boat's intended use, manufacturers will supply the appropriate propeller type. Amongst these are: standard aluminium propellers, stainless steel propellers, propellers allowing a strong thrust in reverse and propellers with variable pitch.

Propeller marking

The propeller marking is generally shown in inches and fractions of inches. The first numbers indicate the diameter; the second ones, the pitch.

Propeller design

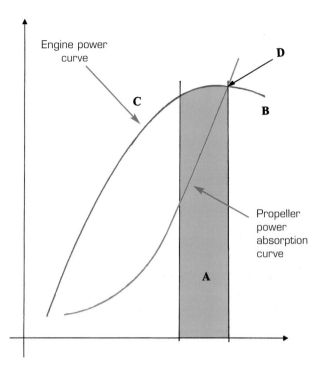

Engine power curve

Propeller power absorption curve

A Optimum engine function zone.
B If the engine exceeds the optimum function zone, the pitch is too small.
C If the engine can't reach the optimum function zone, the pitch is too large.
D The correct propeller is the one that absorbs the total engine power at maximum revs.

Propeller marking

The mark is stamped on the hub or propeller blades. The first number (**14**) shows the propeller diameter; the second (**17**) shows the pitch.

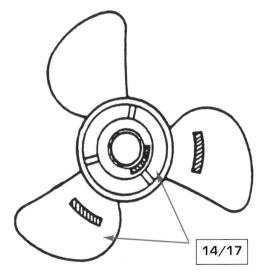

14/17

Pressure distribution on the blades

When the propeller turns, negative pressure is created on the forward face of the blade and positive pressure on the back. The positive pressure pushes the propeller forward; the negative pressure draws it forward.

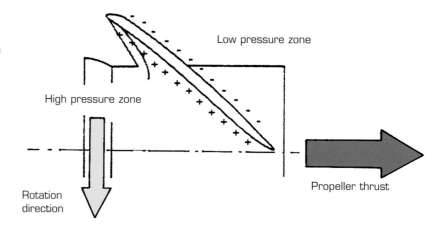

Low pressure zone

High pressure zone

Rotation direction

Propeller thrust

Propeller rotation direction

The direction of rotation can be seen by looking at the propeller from behind the boat.

Clockwise or right-hand

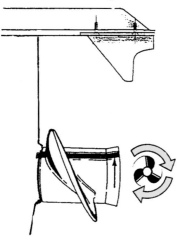

Counter-clockwise or left-hand

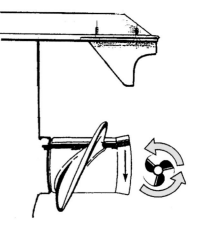

Right rotation

Left rotation

A boat with a single right-hand propeller, in forward motion with the helm at the centre, tends to turn slightly to the left.

A boat with a single left-hand propeller, in forward motion with the helm at the centre, tends to turn slightly to the right.

Special phenomena

Ventilation and cavitation: two phenomena that should not be confused, even though when they appear, the results are the same, ie over-revving and severe loss of thrust.

Ventilation

Ventilation is the result of a loss of thrust caused by exhaust gases or air bubbles being sucked into the propeller. This causes a drop in water pressure on the propeller blades, diminished thrust and increased revs.

Cavitation

This can affect propeller function. The spinning of the propeller blades in the water creates a partial vacuum. As water rushes in to fill the vacuum, tiny air bubbles form, then violently collapse; this reduces the propeller efficiency and also causes pitting on the propeller blades.

Ventilation

Air is sucked into the propeller causing performance loss.

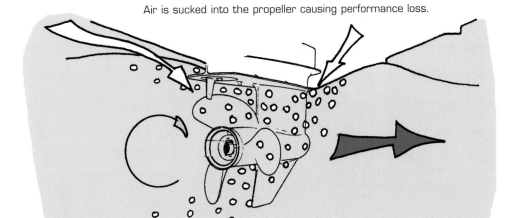

Cavitation

Creation of damaging air bubbles due to a vacuum caused by low pressure on the surfaces of the propeller blades.

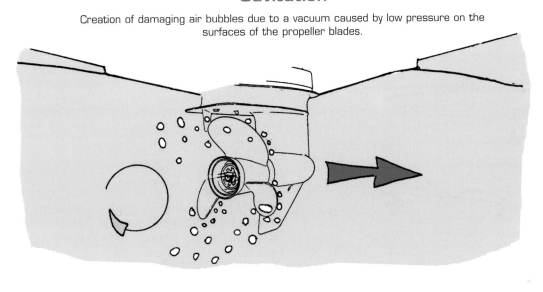

Drive system tilt

Setting the tilt adjustment pin

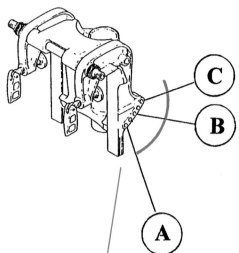

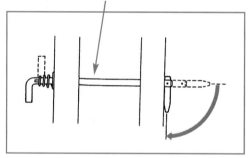

Tilt adjustment pin

Effect on the boat

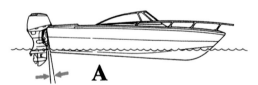

B

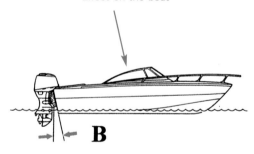

A

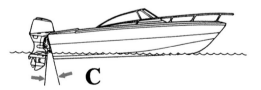

C

Propeller torque

Due to propeller torque on powerful engines, the boat tends to kick or veer sideways. This can be easily corrected by modifying either the load on the boat or the trim tab angle. The trim tab is located on the lower leg of the engine.

Trim tab adjustment

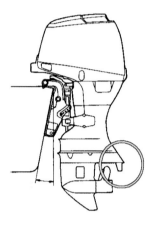

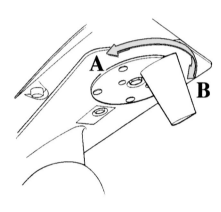

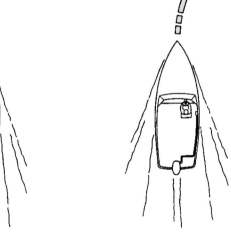

When the boat tends to veer to port:
Turn the trim tab to the left side of the boat (port).
(A on the diagram)

When the boat tends to veer to starboard:
Turn the trim tab to the right side of the boat (starboard)
(B on the diagram)

 CAUTION
The trim tab is also an anode to protect your engine from galvanic corrosion. Never paint the trim tab as this prevents it from doing its job.

Lubrication system

Two-stroke engine lubrication

Because of its design, every two-stroke engine needs to be lubricated by adding oil to the petrol/gasoline. When the fuel/air/oil mix enters the crankcase, it comes into contact with the hot engine casing; the fuel mix gasifies; the oil is deposited, thus lubricating the engine. The amount of oil in the petrol/gasoline is 1% to 2%, depending on the manufacturer.

The mix

What could be more annoying that to have to re-mix oil and petrol/gasoline every time you fill the tank. Although it is a simple operation, over time it becomes tedious. What a hassle! Is it well mixed? Did I put the right amount in? Too much oil and the engine smokes, the spark gets fouled and the engine loses power. Not enough oil and the petrol/gasoline washes the oil film from the cylinders, speeding engine wear and causing the pistons to seize. Over the last few years, oil-injected or automatic mixing systems have appeared on the market, which have made the boat owner's life much easier when it comes to filling up. When the system uses variable oil injection, an additional advantage is that a considerable amount of oil is saved. Not long ago, these devices were only found on high-powered engines. Today, they are offered as standard equipment or as an option on most models.

Examples of these automatic mixing systems are Auto-Blend for Mariner and Mercury, and Accumix for Johnson and Evinrude engines. They guarantee a constant 2% oil mix at any engine revs. Load-based, on-demand variable lubrication devices (0.7 to 2%) are now offered by most manufacturers.

Yamaha Autolube System

The most ingenious automatic lubrication system, shown here on a 6-cylinder model.

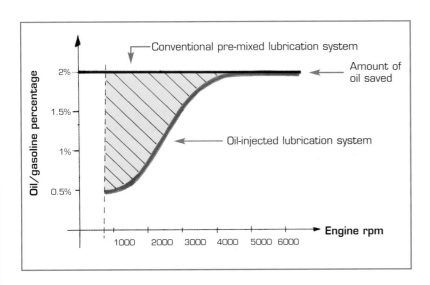

Automatic mixers

These stand-alone devices, such as the Auto-Blend designed for Mariner and Mercury and the Accumix for Johnson and Evinrude engines, were designed for engines without an oil pump. They are easily fitted on most engines, freeing you from the chore of preparing the mix.

Oil injection devices

These lubrication systems, well known in the motorcycle industry, allow a variable mix, from 0.7% at idle to 2% at full throttle.

This on-demand injection system consists of a reservoir and a variable delivery oil pump.

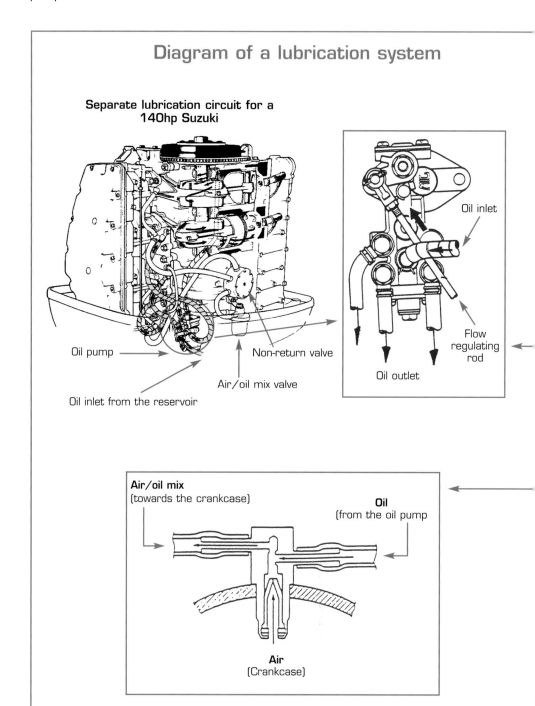

Diagram of a lubrication system

Separate lubrication circuit for a 140hp Suzuki

Oil pump

Oil inlet from the reservoir

Non-return valve

Air/oil mix valve

Oil inlet

Flow regulating rod

Oil outlet

Air/oil mix
(towards the crankcase)

Oil
(from the oil pump

Air
(Crankcase)

How it works

The oil flows from the reservoir, through a filter, towards the oil pump. The oil pump is driven by a gear on the crankshaft.

The pump piston injects a predetermined amount of oil corresponding to the butterfly valve opening and the actual engine load, at a speed governed by the engine revs. The pre-atomised oil is injected behind the carburettor to avoid clogging it.

Audible and visual alarms sound when there is a malfunction, eg insufficient oil reserve, clogged filter, lack of oil.

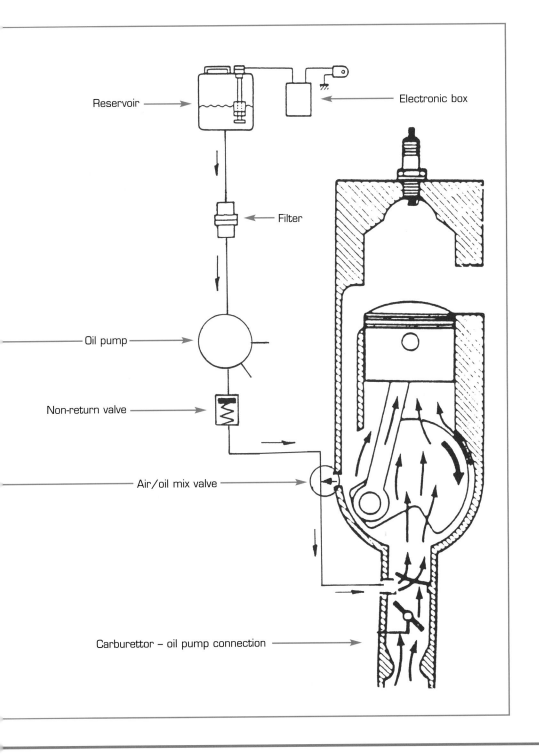

Reservoir

Electronic box

Filter

Oil pump

Non-return valve

Air/oil mix valve

Carburettor – oil pump connection

Oil circuit

Oil pressure
indicator light

Main oil channel

Engine block

Crankcase

Head

Discharge
valve

Oil pump

Oil strainer

Four-stroke engine lubrication system

Four-stroke engines are lubricated by a system of pressurised oil. An oil pump, driven by the camshaft, pumps oil stored in the sump up at a maximum pressure regulated by a discharge valve and pushes it through the various channels that feed the crankshaft journals, the camshaft journals and the rocker arms. Then gravity causes the oil to fall into the sump through channels for that purpose. An oil filter on the circuit provides oil filtration.

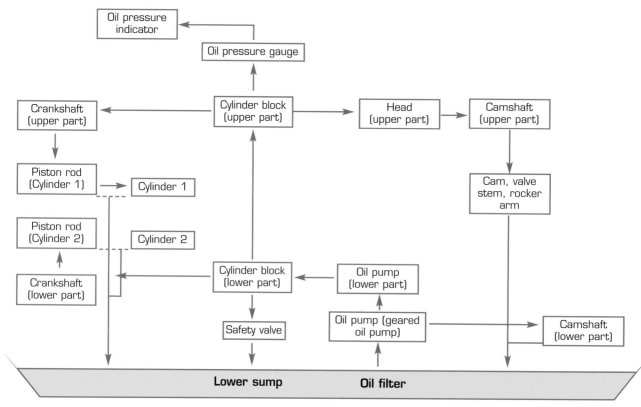

Oil pressure
indicator

Oil pressure gauge

| Crankshaft (upper part) | Cylinder block (upper part) | Head (upper part) | Camshaft (upper part) |

Piston rod (Cylinder 1) → Cylinder 1

Cam, valve stem, rocker arm

Piston rod (Cylinder 2) — Cylinder 2

Crankshaft (lower part)

| Cylinder block (lower part) | Oil pump (lower part) | Camshaft (lower part) |

Safety valve

Oil pump (geared oil pump)

Lower sump   **Oil filter**

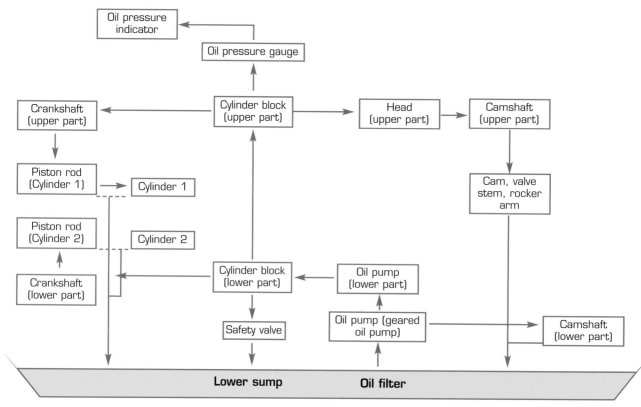

MAINTENANCE

DESPITE THE MANY IMPROVEMENTS in outboards over the past few years, breakdowns still happen. How do you prepare yourself? By performing careful checks before starting, as well as while the outboard is running, and by carefully following the recommended maintenance schedule.

Depending on how careful or careless you are, the time you spend boating will be either worry free or full of problems. There is a worksheet for each maintenance task that shows you the tools, materials and time usually required. Each task is colour-coded as either *simple*, *technical* or *complex* depending on how much skill is needed (see page 66).

SYSTEMS CHECKS

Checking the engine condition properly will allow you to have a fairly exact idea of its degree of efficiency and general condition.

The outboard is essentially divided into two parts:

1   The peripheral equipment that controls the engine function, ie the fuel, cooling and ignition systems.
2   The mechanical ensemble, ie the engine itself.

All of these elements are subject to wear, which means that they will have to be adjusted or replaced after a certain number of hours.

We can easily understand that the engine block requires less maintenance than the peripheral systems, assuming that we follow the manufacturer's recommendations regarding running conditions (warm up, maximum revs), the choice of oil used in the fuel mixture for two-stroke engines, and the oil quality and the frequency of oil changes for four-stroke engines.

Nonetheless, the engine's constituent parts are subject to a certain amount of wear. As the engine is used, some parts become loose, mechanical noise increases; engine performance decreases and fuel consumption rises. The first symptoms of malfunction appear: various leaks, poor starting, overheating.

There are at least five checks that you can perform to monitor your engine's condition.

1   Check the instruments (oil level and pressure for four-stroke engines). Reading the oil pressure on a four-stroke engine tells you how worn the engine is.
2   Listen for unusual mechanical noises. Using a stethoscope will help you precisely locate the origin of a noise.
3   Check the cooling system. This check will tell you the efficiency and quality of engine cooling.
4   Check the fuel system. Checking the fuel circuit for leaks, and checking the richness of the fuel mix by looking at the spark plugs, will tell you the quality of combustion in the heart of the combustion chamber.
5   Check the compression. This procedure allows you to check the head gasket as well as the crankcase seals on a two-stroke engine.

Checking the instruments

Your engine is equipped with different instruments: tachometer, water pressure gauge, temperature indicator, fuel gauge. Some are connected to audible alarms, notably the cooling and lubrication systems.

The purpose of all of these instruments is to allow you to continually monitor your engine's functioning, its revs, and its operating temperature.

The tachometer

The tachometer shows the engine revs in revolutions per minute (rpm). You shouldn't constantly run the engine at maximum revs. Always have a margin of safety. Choose your cruising speed bearing in mind your engine's capability. Optimum fuel efficiency is at your engine's maximum torque.

As a general rule, once your engine is warm, its revs should be steady and should stabilise at idle between 850 and 950rpm and when in gear between 650 and 800rpm.

High revs at idle will damage your gears when going into forward or reverse. Also, an irregular idle should prompt you to perform more in-depth checks.

Instrument panel

A safety measure: the most complete instrument panel possible
Reduced to the bare minimum on small outboards, the instrument panel acquires more significance on higher powered engines. Below is a sample instrument panel with the instruments needed to ensure proper functioning of a boat and its engine.

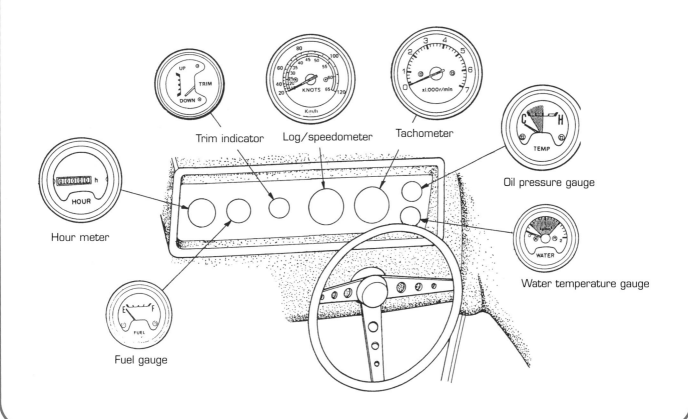

Trim indicator   Log/speedometer   Tachometer

Oil pressure gauge

Hour meter

Water temperature gauge

Fuel gauge

Water temperature and pressure gauges

These gauges are essential for monitoring your engine temperature as well as the water pressure in the circuit.

A sudden rise in water temperature or a drop in water pressure should alert you. Stop the engine immediately; check that the cooling water inlets are not clogged.

If the condition persists, a further check of the cooling system is needed.

Audible alarm/buzzer

The alarm sounds when the engine starts to overheat or when the oil level and lubrication are insufficient. It is vital to stop the engine and check the lubricating oil level and the cooling water inlets.

On some engines, when the alarm sounds, a protection circuit limits the engine revs. The alarm continues to sound until the temperature around the thermo-couplers goes below the alarm level.

Be careful: This type of instrument will not tell you the cause of the fault. You will have to closely examine the different elements of the cooling and lubrication circuits to identify the problem.

Noises and their probable cause

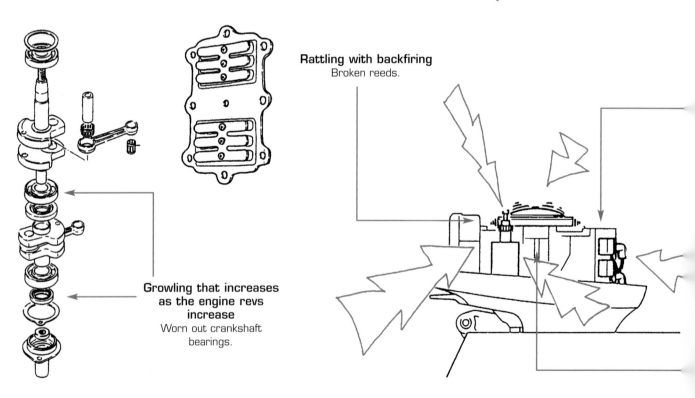

Rattling with backfiring
Broken reeds.

Growling that increases as the engine revs increase
Worn out crankshaft bearings.

When the engine is running, listen carefully and try to track down the suspicious noise. Banging, rattling and grinding should alert you. If you're unsure of the source of the noise, use a mechanic's stethoscope. This tool (which is similar to those used by doctors except that the amplification is lower) will allow you to locate the source of a mechanical noise with great precision.

> **If you don't have a stethoscope, you can use a wooden stick or a plastic tube. They will transmit the various mechanical noises, although much less audibly.**

As a general rule

A metallic noise from the cylinder area means that the piston rings are damaged.

   A continual rumbling that increases with engine revs should point you to the crankshaft bearings.

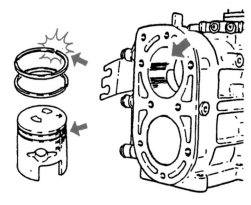

Whistling by the engine head
Head gasket leaks.

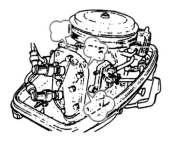

Backfire at full throttle
Wrong spark plugs, cracked or broken insulator, bad timing adjustment.

Metallic click-clack near the cylinders
Play in the piston rings or broken piston rings.

Clanking on acceleration
Play in the piston pin.
Worn out piston rod bearings.

The aim of your outboard's cooling system is to dissipate the excess heat produced by the combustion of the fuel/air mix in the cylinders and to regulate the engine temperature. So it is easy to understand that even a momentary absence of cooling can cause irreparable damage to the engine.

Outboards have a water cooling system. A pump circulates the water through the engine; a thermostat regulates the temperature.

A tell-tale jet below the engine shows if the cooling system is functioning properly. You should pay attention to this as it will tell you how well the water is circulating inside your engine and can help you locate the source of a cooling system problem.

Locate the following on your engine:

◆ the cooling water inlets
◆ tell-tale jet

With the engine running, check the jet flow and temperature at different revs

As we saw in the section on engine cooling, remember:

◆ A good flow of tepid water means there is no problem; the engine is functioning properly.
◆ If the flow is good but the water temperature is over 60°C, the pump is functioning normally, but the thermostat is stuck in the 'closed' position.
◆ If the flow is weak and too hot, check the water pump.
◆ If the flow is weak and cold, the thermostat is stuck wide open or someone has taken it off. Remove and check the thermostat.
◆ An intermittent flow or absence of flow from the tell-tale jet, despite normal engine temperature, means that the tell-tale circuit is clogged. Blow compressed air into the tell-tale hole. If that doesn't clear it, the whole circuit has to be checked further and cleaned.
◆ No flow, a cloud of smoke coming out of the tell-tale. Stop the engine. Check the entire system. It is likely that all the parts are worn and that the cooling channels are obstructed.

> **IMPORTANT**
> **Before testing the engine, make sure the cooling water inlets are not clogged.**

Possible causes of problems

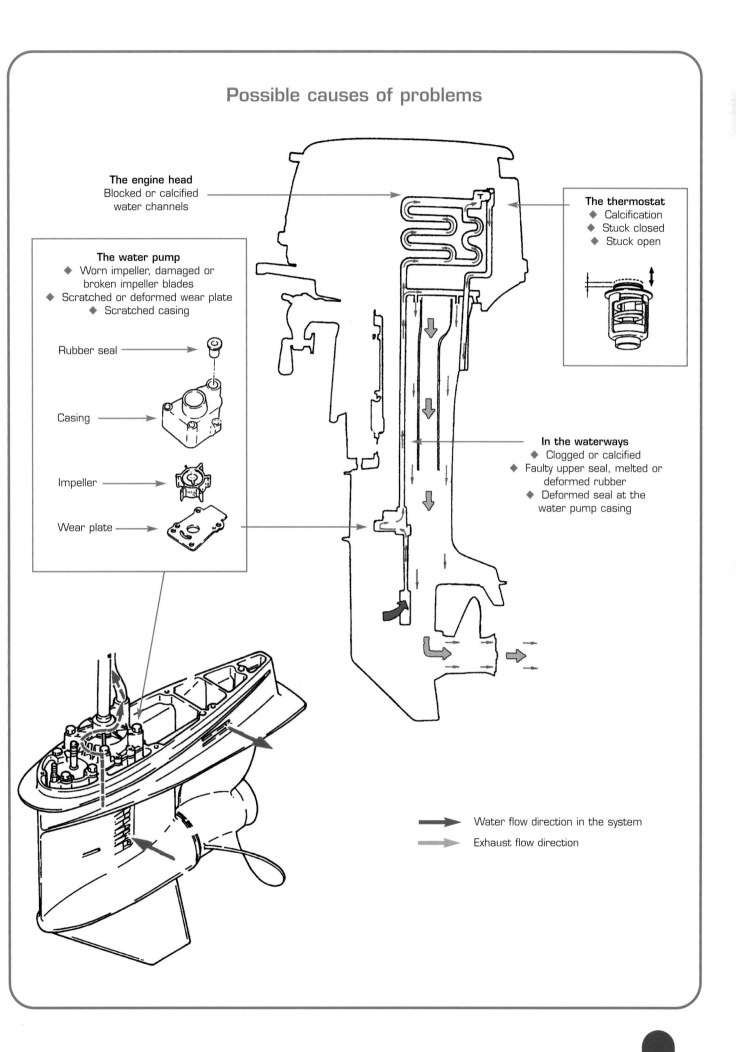

The engine head
Blocked or calcified
water channels

The thermostat
◆ Calcification
◆ Stuck closed
◆ Stuck open

The water pump
◆ Worn impeller, damaged or
broken impeller blades
◆ Scratched or deformed wear plate
◆ Scratched casing

Rubber seal

Casing

Impeller

Wear plate

In the waterways
◆ Clogged or calcified
◆ Faulty upper seal, melted or
deformed rubber
◆ Deformed seal at the
water pump casing

→ Water flow direction in the system

→ Exhaust flow direction

CHECKING THE FUEL SYSTEM

Any defect in the fuel system causes engine problems: misfiring, loss of power, irregular running, poor starting, loss of acceleration, stalls, etc. To prevent these problems, it is important to check, tune and fix certain things, the first being to check the fuel system for leaks. This can be done as you start the engine and requires no special tools.

How to check the fuel tank for leaks

◆ If the fuel tank has an air inlet, check if it is clear.
◆ Check the connections on the fuel line.
◆ Squeeze the priming valve on the fuel line until it becomes hard.
◆ Keep the pressure on the priming valve for a few seconds.
◆ Visually check each connection for leaks all the way to the carburettor.
◆ Any leak in the system inevitably allows fuel to escape. The leak causes the engine to malfunction because air is sucked into the system as the engine runs.

Before starting the engine, perform a quick check of the fuel system.

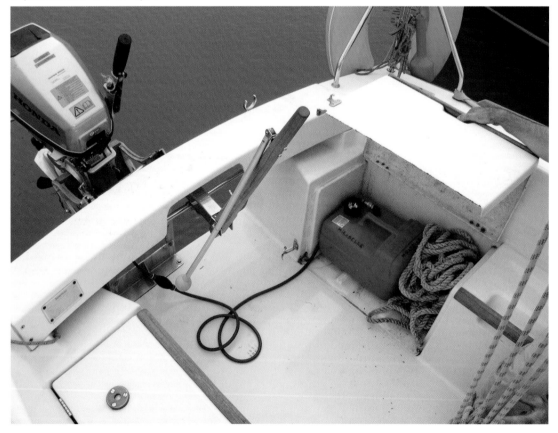

Checking the fuel system

Carefully check all points marked with a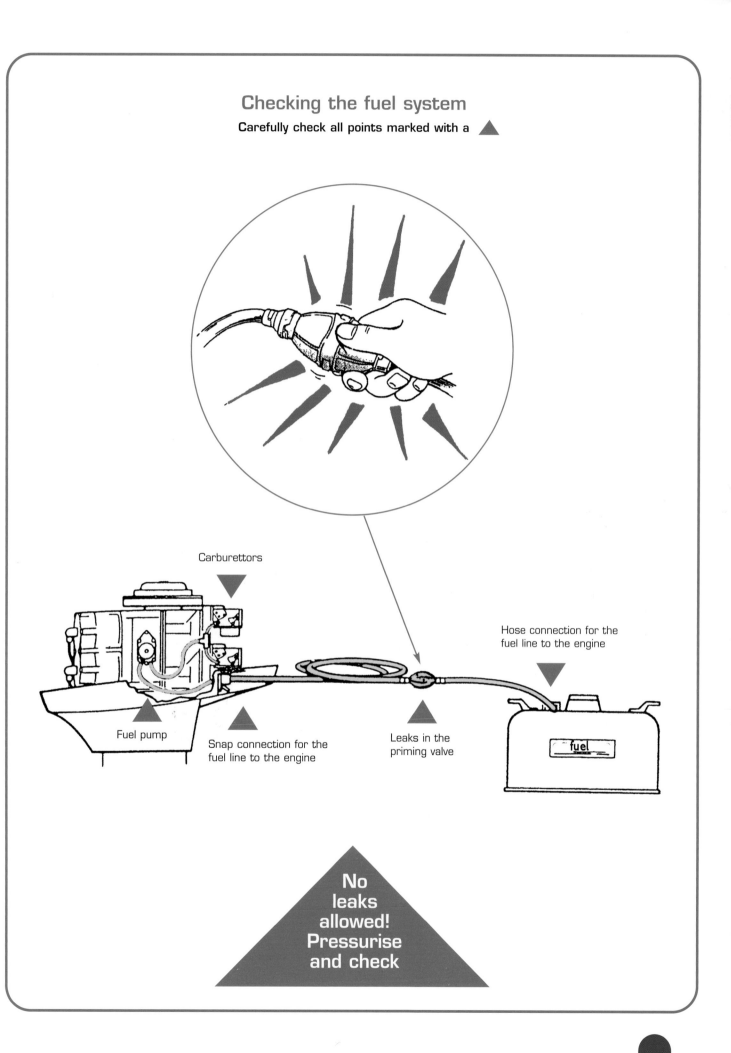

Carburettors

Hose connection for the
fuel line to the engine

Fuel pump

Snap connection for the
fuel line to the engine

Leaks in the
priming valve

fuel

No
leaks
allowed!
Pressurise
and check

Regular maintenance of your outboard engine is the first line of defence against damage caused by wear and tear.

Follow the maintenance schedule in your outboard engine manual.

Certain manuals delivered with the engine are very detailed; others are very basic. If you don't have a manual for your engine, this chapter provides a table showing what should be checked and when, as well as a description of the maintenance operation to be performed. You will have to adapt the operations described in the instructions to the engine in question (four- or two-stroke, power, electric starter, hydraulic trim/tilt, etc).

Adjusting the idling rate.

 IMPORTANT

Your outboard engine was designed to run with its original parts. Use only the manufacturer's parts to ensure proper performance and warranty. Also, follow the recommended maintenance schedule.

Your safety and your engine's longevity depend on this.

Maintenance schedule

Points to check	What to do	When
Fuel filter	Check – clean	Every 6 months or every 100 hours
Fuel tank	Check – clean	Every year
Fuel lines	Check for leaks	Every use
Spark plugs	Check – clean Replace	Every 50 hours Every 200 hours or every year
Electrical system	Check – renew	Every 100 hours or every year
Kill switch	Check – renew	Each use, if necessary
Battery	Check – fill as needed	Every 100 hours or every year
Engine oil	Change	Every 100 hours or every year and every 50 hours for small engines
Oil pump	Check – adjust as needed	Every year or 100 hours
Propeller	Check	Regularly
Lubrication points	Lubricate	Every 100 hours
Anodes	Check – renew	Every 200 hours or every year
Gearbox oil	Check – renew	Every 100 hours or every year
Water pump – thermostat	Check – replace if necessary	Every 200 hours or or every year
Linkage	Check – adjust as needed	Every 200 hours or every year

HANDS-ON TASKS

Basic tool kit

As an outboard user, you should always have a set of tools on hand for doing simple jobs. It is what I call the basic tool kit; it includes:

◆ **Spanners**
Flat and mixed. These are the essentials in any tool kit. **Caution**: When you work on a engine built in the USA (Evinrude, Johnson, Mercury, Force), use spanners sized in inches and fractions of inches, not metric.

◆ **Screwdrivers**
Include at least two flat screwdrivers and two cross-point screwdrivers.

◆ **Pliers**
General purpose, cutting, or vice grip – each has a specific application.

◆ **Hammer**
Brutal, but indispensable.

◆ **Specialised tools**
A spark plug spanner and its set of feeler gauges are indispensable, as is a wire brush.
The multimeter, useful in many cases for checking electrical circuits, is also one of the tools you can't do without.

◆ **Useful products and accessories**
Keep on hand a spray can of penetrating oil, marine grease and an oil can.
Maintenance is a dirty job, so you'd best have some rags and liquid soap to wash your hands. Good luck!

First you have to invest in a basic set of good tools (see box). Then, you will have to spend some of your spare time studying the different systems you're going to work on (ignition, cooling, fuel).

If you are a beginner or if you have any doubts, it would be a good idea to review the basic concepts described in the beginning of this book before starting any work.

One rule to remember: always choose the best quality tools; they will obviously be more expensive, but so much more reliable. If I had to name one brand, I would say Snapon. But in any case, remember this old adage: good tools make a good worker.

This book isn't intended to replace the manuals for each engine, but the worksheets will guide you, and inform you how to approach the task and what to watch out for before starting each job.

USING THE WORKSHEETS

The degree of difficulty of a task is shown by its colour and description

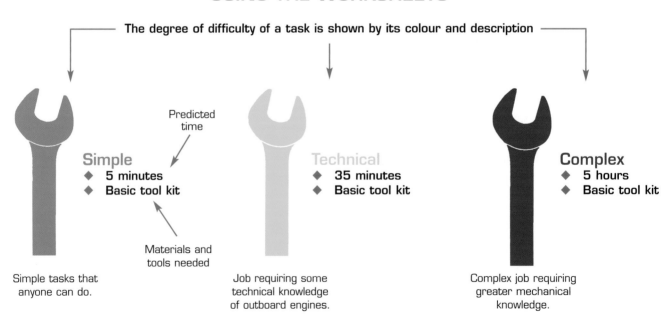

Predicted time

Simple
◆ **5 minutes**
◆ **Basic tool kit**

Materials and tools needed

Simple tasks that anyone can do.

Technical
◆ **35 minutes**
◆ **Basic tool kit**

Job requiring some technical knowledge of outboard engines.

Complex
◆ **5 hours**
◆ **Basic tool kit**

Complex job requiring greater mechanical knowledge.

Checking the spark plugs

Simple

- 10–20 minutes, depending on the number of spark plugs
- spark plug spanner or special socket, a set of feeler gauges, a small, soft brush

Spark plugs are an essential part of the electrical system. They ignite the fuel/air mixture. Therefore, it is important to check them regularly.

The appearance of a spark plug tells you about the condition of the engine and its state of adjustment. Significant deposits, a deformed or worn out electrode, a chipped or broken insulator – these are all indicators that give you valuable information about your engine:

- ◆ Fuel mix richness
- ◆ Timing
- ◆ Condition of the rings

> **! IMPORTANT**
> **The spark plugs should be checked while the engine is warm from running under normal conditions.**

● Remove the spark plugs

If you have never done this, refer to the worksheet 'Replace and adjust the spark plugs'.

● What to look for

A slightly damp, blackish deposit may mean:
- ◆ A fuel mixture that is too rich
- ◆ Too much use of the choke
- ◆ Running the engine too slowly for a long period
- ◆ A delay in the timing
- ◆ A spark plug with too high a temperature rating (too 'cool')

A whitish or light grey colour is caused mainly by:
- ◆ A fuel mixture that is too lean
- ◆ A spark plug with too low a temperature rating (too 'hot')

When the spark plug insulator is light brown and shows slight wear on the electrodes:
- ◆ The thermal rating is correct
- ◆ The adjustments are correct
- ◆ The wear is normal

Normal
The insulator is light brown in colour. Its thermal rating is appropriate.

Worn out electrodes
This spark plug shows signs of over 200 hours of use. Replace it with a spark plug of the same type.

Dry, black deposits
This is usually a sign that the fuel mixture is too rich, but before adjusting your carburettor check to see if the engine compression is a bit weak.

Oily deposits
Incomplete oil combustion, which could be caused by too much oil in the fuel mixture, or a spark plug or engine temperature that is too cool.

Burnt electrodes
Overheated spark plug. Incorrect timing adjustment, defective cooling, inappropriate thermal rating for the application. Use a 'cooler' spark plug.

Simple

- 10–30 minutes, depending on the number of spark plugs
- Spark plug spanner or special socket, set of feeler gauges, new spark plugs

In principle, spark plugs should be changed every 200 hours or whenever the engine is accidentally flooded. Adjustment and cleaning should be done every 50 hours.

● Preparation

Get the manufacturer's recommended brand and type of spark plug.

● Removal

- ◆ Let the engine cool. Remove the cover.
- ◆ If your engine has several cylinders, note the leads and change the spark plugs one at a time (so as not to mix up the leads).
- ◆ Disconnect the spark plug leads by pulling on the insulator cap and twisting slightly. Avoid pulling directly on the wire.
- ◆ Unscrew the spark plug with a 16 or 21mm socket, making sure that you keep the axis aligned.

> ## IMPORTANT
> Make sure that the area immediately around the spark plug is very clean before you remove it. Clean with a paint brush and petrol. Blow or wipe it dry.

● Inspection

If the wear on the electrode is severe or if the carbon deposits are excessive, the spark plug(s) must be replaced.

As a general rule, it is recommended that the spark plugs be replaced every 200 hours. But a clean-up can get you by, or prolong its life.

● Adjustment

Adjust the electrode gap by carefully bending the external electrode, taking care not to put pressure on the central electrode.

Check the gap by using the correct feeler gauge. The correct gap measurement for your engine can be found in your manual. Generally, it is 1mm for electronic ignition engines and 0.7mm for older engines equipped with points.

● Reinstallation

Before reinstalling the spark plugs, check the specifications, the thread reach (short or long) and the condition of the spark plug seat in the engine head.

SPECIFICATION TABLE

Make	Temperature rating	NGK	Champion	AC	Bosch	KLG	Lodge	ND
				47FF	W45T1			
	Hot	B-4H	L14			F20	BN	W14FS
				46FF	W95T1			
		B-5HS	L10, L90	45FF			2H, C14	W16FS
					W145T1	F50		
				45F			CC14, CN	
		B-6HS	L88, L86	44F	W175T1	F70		W20FS
			L87Y	42F				
						F75	H14, HN	
			L7, L7J	44FF, M43FF			HBN	
			L81, L4J		W225T1			
							HN-14	
		B-7HS	L5, L78	42FF	W240T1	F80	2HN, 3HN	W22FS
		B-8HS		MC41F				W24FS
		(B-8HS10)		M42FF				
			L77J					
					W260T1	F100	HH14	
	Cold	B-9HS						W27FS
					W280T1			

(Make column illustration labels: 0.5–0.6mm, 12.7mm, 14mm)

Hand-tighten the spark plug, making sure it is centred, until it makes contact with the head. Do not force it; it may not be straight. Then tighten it a quarter of a turn with the spanner/spark plug socket. *Be careful*: excessive tightening may damage the head. If you use a torque wrench, set it at 20–25Nm.

For spark plugs with a conical seat (without an external washer), tighten 15Nm or 1/16 of a turn.

Reconnect the spark plug leads correctly.

Tightening and loosening the spark plugs

The spark plug has to be screwed in by hand before using a spark plug spanner.

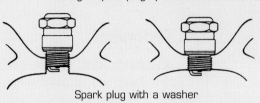

Spark plug with a washer

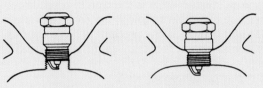

Spark plug with conical seat

Tighten 1/4 turn for spark plugs with a washer

Tighten 1/16 of a turn for spark plugs with a conical seat

1 Before removing the spark plugs, clean around the spark plug socket with a brush and petrol.

2 Blow and dry.

4 Cleaning with a soft wire brush removes deposits and other combustion residues.

3 Loosen the spark plug making sure you keep the spanner well aligned.

5 Follow the manufacturer's specifications for the gap.

Replacing the starter cord

Technical

- From 30 minutes to 1 hour, depending on the type of starter unit
- Basic tool kit, starter cord

A broken starter – this is the type of breakdown that should be avoidable so check the cord regularly for wear and chafing. If the problem is more serious, ie if the starter spring is broken, give the repair job to a professional. Replacing the starter cord is fairly easy, but replacing the spring isn't.

Removal

◆ Remove the starter unit.
◆ Completely unwind the worn cord, block the drum, then pull the cord out by pulling on the knotted end.
◆ If the cord is broken, the recoil spring is unwound. Turn the drum anticlockwise. This will tighten the spring.
◆ After the spring is wound tight, block the drum leaving it so that the hole for the cord is in line with the cord's exit on the starter unit.

Change

◆ Prepare a new cord with a knot at the end.
◆ Slide the new cord into the drum's hole, then through the starter body. Pull on the cord so the knot is in contact with the drum.
◆ Unlock and slowly release the drum. The cord should wind itself around it. Pull the cord through the pull handle and fasten it.

Test

◆ Before refitting, try the starter unit several times.
◆ The pull handle has to come back to its resting place each time.

Refitting the starter

Remount the starter unit onto the engine.

Lock the drum. A mole wrench or vice grip is good for this.

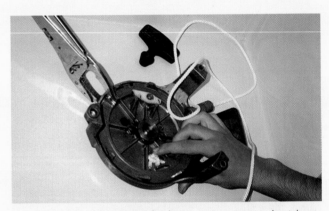

Install the new cord, remembering to put a stopper knot in the drum end.

Pull on the cord several times before refitting.

Dismantle, check and refit the thermostat

Complex

- 30 minutes to 1 hour, depending on how accessible the thermostat is.
- Basic tool kit, saucepan or baby bottle heater, thermometer.

The thermostat's job is to regulate the engine temperature in accordance with the manufacturer's specifications. Engine overheating is often caused by the closure or partial opening of the thermostat. A very cold engine means that the thermostat valve stays open or opens too early.

This test will allow you to check the opening and closing of the thermostat and to determine if the water lines are clogged.

● Find the thermostat

It is usually located on the upper part of the engine, close to the head.

● Remove

Remove the thermostat cover. Pull out the thermostat.

● Examine

Examine the thermostat for signs of deterioration. If it is damaged or if it is open when cold, replace it.

Heavy deposits in the water lines are a sign that the engine is running too hot and there is poor water circulation.

● Clean

Clean the circuit inside the engine block. Eliminate all traces of salt, deposits and corrosion with a brush and water.

If the circuit is completely clogged up, the entire engine must be descaled and desalted to re-establish the cooling water circuit.

● Check

Submerge the thermostat in a container full of water. Gradually heat the water.

Note at what temperature the thermostat opens and the degree of opening.

If it doesn't open, or the temperature at which it opens and the size of the opening don't match the manufacturer's specifications, replace the thermostat.

● Refit

Scrape off the old seal. Put on a new seal. Refit the thermostat.

> **IMPORTANT**
> **The thermostat valve should open at the temperature stamped on the thermostat.**

Find the thermostat and remove its cover.

Remove the thermostat. Scrape clean, then blow off the deposits that have accumulated (salt, scale, etc).

Check the thermostat temperature

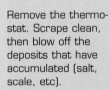
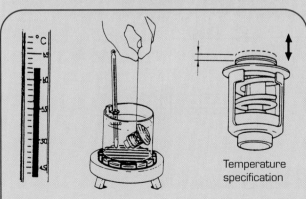

Temperature specification

Check the thermostat opening by heating the water.

Changing the gearbox oil

Technical

- 30 minutes
- Large scewdriver, rag, small container, tube of gearbox oil

You should change the gearbox oil at the end of each season or every 100 hours.

Preparation

Get a small container, then unscrew the upper gearcase oil plug, located above the oil level mark.

Drain the gearbox

Remove the upper vent plug. Remove the lower plug. Allow oil to drain.

Refill the gearbox

Using a tube of gearbox oil, refill the gearbox through the lower oil plug opening. Squeeze the tube until the oil pours out of the upper vent plug opening, then close it. Promptly remove the tube with one hand and with the other, have the plug ready to close the lower plug opening.
Firmly tighten the plugs.

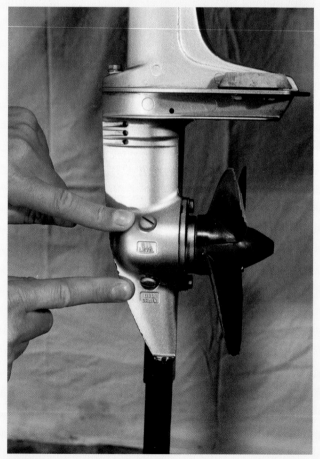

1 Note the upper and lower plugs.

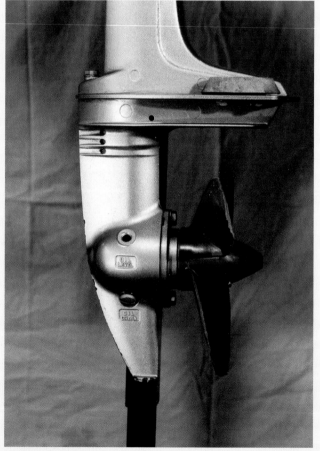

2 First remove the upper plug to allow air in.

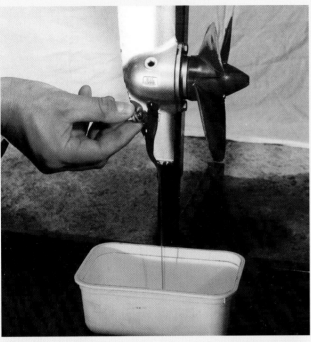

3 Remove the gearbox oil plug. Note the oil colour. A whitish colour is a sign of water in the oil.

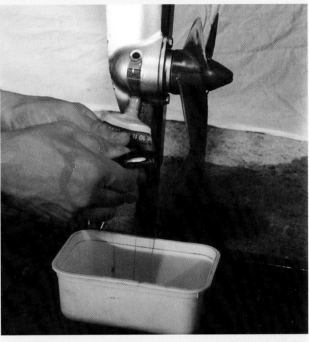

4 Fill the gearbox through the lower oil plug opening until there are no air bubbles coming out and the oil flows out of the upper opening.

5 Screw in the upper plug, remove the tube and screw in the lower plug.

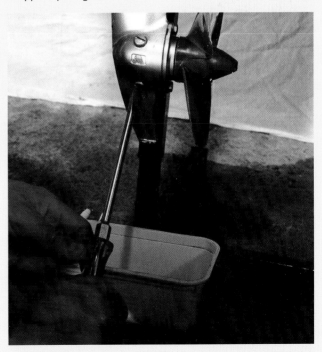

6 Check that the plugs are screwed in tightly. Clean.

Changing the engine oil

Simple

● 1 hour

● Large screwdriver or spanner, oil filter spanner, bucket to catch the oil, oil, oil filter (if necessary), degreaser, rag

Unlike their two-stroke counterparts, four-stroke engines need specific maintenance. The different engine components are lubricated by a pressurised oil system.

Despite the improvements in lubricants in the last few years, oil degrades during engine use. In fact, when oil is subjected to extreme heat and wear over time, it becomes filled with metallic impurities and combustion residues. That is what gives it its blackish colour. On a four-stroke engine, other than checking the oil level with the dipstick, changing the oil is the most important thing to do.

This job can be done by anyone because it doesn't require any particular mechanical knowledge, or any specialised tools. *Beware*: changing the oil is a necessary but very messy job if you don't take some precautions. Make sure you work in a methodical way.

● Maintenance schedule

Follow the manufacturer's recommended maintenance schedule.

In general, every 100 hours, annually, or every 50 hours for small engines (2–4hp).

● Preparation and precautions

First you will need to:
◆ Know how much oil you will need – check the manual. If you are changing the filter, allow an extra 0.15 to 0.25 litres. Small engines don't have an oil filter but are equipped with an oil strainer.

Assemble the tools and products needed:
◆ Large screwdriver or spanner, depending on the engine model
◆ Oil filter spanner
◆ Bucket to catch the oil
◆ Oil
◆ Oil filter (if there is one fitted)
◆ Degreaser
◆ Rag

Find the different points where you'll be working:
◆ Dipstick
◆ Filler hole
◆ Oil filter

The filler hole is often also the drain hole.

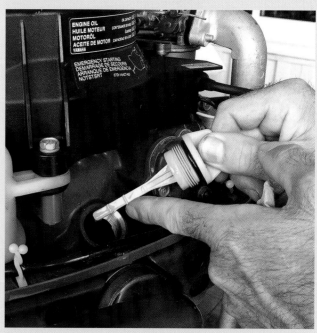

Check the oil level in a four-stroke engine regularly by unscrewing the oil cap and checking the dipstick.

FOUR STROKE

Some engines have an indicator light that displays the oil level without the need to open the cover, as seen on this Johnson engine.

● How to do it

Always change the oil when the engine is warm to ensure that the oil drains quickly and completely.

You can get the engine up to temperature in a tank, on the water, by using the flushing system provided by some manufacturers, or by using a flushing device (muffs and a hose pipe) that fits onto the water inlet. In this last case, you must adjust the hose pipe water flow. One way to tell if it is correct is if the stream of water from the tell-tale is normal.

> ✸ If you use a tank, muffs, or the flushing system and your engine doesn't have gears (2hp, 3hp), remove the propeller to avoid accidents and water splashes.

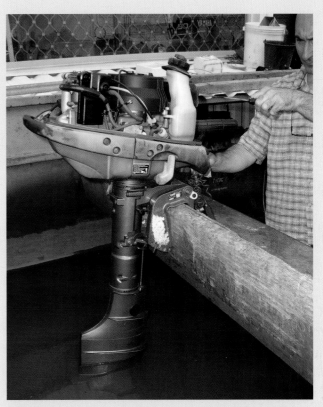

Running the engine to temperature in a tank allows you to safely rinse the water circuits and to check other things, such as the starter, idle, etc.

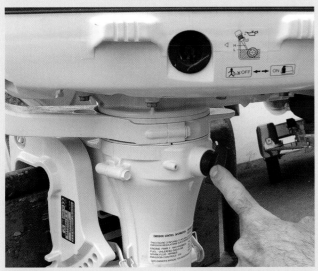

Some manufacturers offer a very practical flushing system on their engines.

Once the engine is running, check that the water flow from the tell-tale is normal. The temperature must be below 55°C.

The procedure is practically identical for every engine. The differences will be in the location of the drain holes, oil filter and the quantity of oil needed.

Note the location, of the oil drain plug (left) on a Yamaha 4hp and (right) on an Evinrude Johnson 4hp.

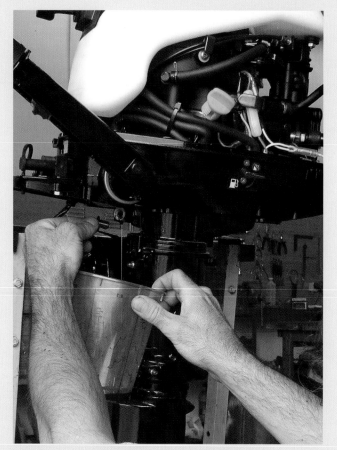

To help draining, remember to remove the oil cap.

Start draining the oil a few minutes after you have turned off the engine.

◆ Begin by removing the oil dipstick or the oil cap to allow the oil to drain.
◆ Unscrew the oil drain plug. Remember not to lose the washer seal.
◆ Catch the oil in the bucket.
◆ Wait until the oil has drained.

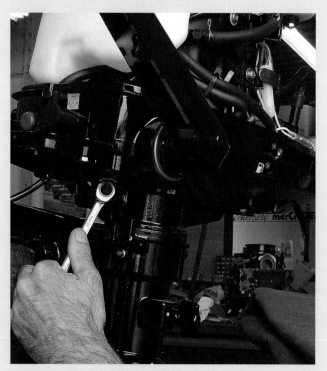

Use a box-end spanner to remove the drain plug.

Check the oil for metallic particles and water.

Meanwhile, if your engine has an oil filter, remove it. Slide a plastic bag under the filter to avoid a mess. Remember to clean the filter seat. A quick wipe with a rag is enough. Screw in the new filter, making sure that you first lubricate its seal with a smear of oil. Screw in the new filter by hand until it touches the block, then tighten an extra $^1/_2$ to $^3/_4$ turn.

 Certain engines don't have a separate oil filter. It is integrated into the drain plug. Clean it with petrol and don't hesitate to change it if there is any doubt about its condition.

◆ Replace the drain plug or screw after changing the washer seal, first screwing it in by hand so as not to damage the threads. Tighten the drain plug or screw.
◆ Put a funnel in the filler hole, then fill the engine oil sump until it reaches the maximum level on the dipstick. (Some engines have a convenient window to see the level.)
◆ Replace the oil cap/dipstick.
◆ Run the engine at idle for a few minutes.
◆ Stop the engine.
◆ Remove the dipstick and check the oil level 2–3 minutes after you have stopped the engine. Top up if necessary.

! IMPORTANT
Make sure that the oil pressure light or alarm goes off after running the engine for a few seconds and that there are no leaks around the filter or drain plug/screw.

Caution: don't pollute the environment!

Your used engine oil is an environmental pollutant. Dispose of your oil properly. Take it to an appropriate disposal point, which is usually provided in all marinas.

● Which brand of oil do you use?

Each manufacturer or dealer usually recommends a certain brand of oil. During the warranty period, you should follow their recommendation. After that, you can change brands, but make sure you use the same quality.

● What type of oil do you use?

A good quality oil ensures proper engine function. So never economise and always follow the manufacturer's recommended viscosity for your motoring conditions.

Use a funnel to avoid spills. Pour the new oil in slowly to aid flow and prevent overfilling.

Oil specifications are clearly stated on the manufacturer's labels. Follow the engine manufacturer's recommendations.

Removing the lower leg

Complex

- From 15 to 30 minutes, depending on engine type
- Basic tool kit, marine grease

● Why should you do this?

You have to remove the lower leg to :

- ◆ Check the water pump
- ◆ Service the linkage/gearbox

For small engines, remove the outboard and attach it to a stand. If you don't have a purpose-built stand, put the engine in a vertical position by clamping it to a board held in a vice.

   For outboards bolted onto a stern plate, tilt the outboard at a 45° angle.

● Removal

- ◆ Remove the kill cord (lanyard emergency stop switch) or disconnect the spark plugs. Uncouple the forward and reverse linkages. *Caution*: study the linkage system before removing it, as each manufacturer has its own system.
- ◆ Unfasten the lower leg screws. If the trim tab has to be removed, note its position so you can put it back the way it was.
- ◆ Detach and remove the lower leg.
- ◆ Do not change the linkage position.

> ## ! IMPORTANT
> **Before remounting the lower leg you should grease the drive shaft splines with waterproof grease.**

● Reassembly

- ◆ Put the lower leg in place. Slide the water tube into position.
- ◆ Make sure the gearbox control goes through where it should.
- ◆ Slightly rotate the flywheel (with the kill cord off) to ensure proper alignment of the drive shaft splines.
- ◆ Tighten the screws a bit at a time.
- ◆ Check and adjust the gearbox linkage if needed.
- ◆ With the control handle in neutral, the propeller should spin freely by hand.

Note
- ◆ Gearbox screws
- ◆ Type of gear linkage system

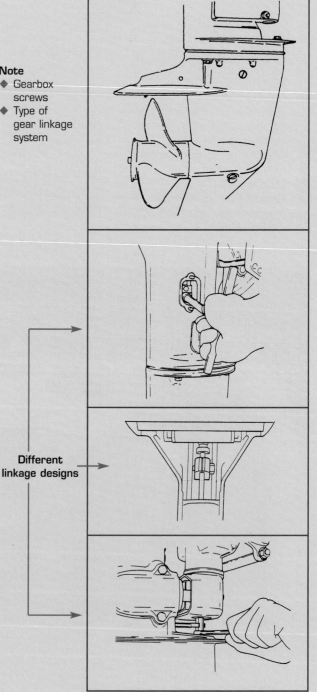

Different linkage designs

1 First disconnect the gearbox linkage, then unscrew the bolts holding the lower leg in place.

2 Remove the lower leg. If needed, give it a tap to loosen it.

4 To replace, lift the lower leg into place, then tighten the screws a bit at a time.

3 Before reassembling the lower leg, remember to clean and grease the splines on the drive shaft.

5 Check the gearbox linkage system. In neutral, the propeller should spin freely. If necessary adjust the linkage connection.

Checking the water pump

Technical

- 15 minutes
- Basic tool kit, water pump repair kit

You should check the water pump every 200 hours or annually to make sure the cooling system is working well.

This job involves removing the lower leg to expose the water pump, which is located above the gearbox in the lower leg.

 On some small engines (2hp), the water pump is on the propeller shaft. In this case, you don't need to dismantle the lower leg to do this job.

Removal

- Remove the lower leg (refer to the worksheet 'Removing the lower leg').
- Remove the screws on the water pump casing.
- Lift the case and slide it up along the drive shaft.
- Remove the impeller in the same way.
- Remember not to lose the little locking key.
- Remove the wear plate under the impeller.

Cleaning

Clean the following: water intake grids, the area under the wear plate, the wear plate, the water pump casing, the seats, the drive shaft.

 If your engine has removable intake grids, don't hesitate to remove them. Note the flow direction. Clean the intakes with a paint brush or soft brush. Reinstall them making sure they are in the same position.

Inspection

Check the impeller blades for wear and suppleness. Pull on them gently to see if they are separating from the hub.

Check the wear plate and pump casing. They should be smooth and in perfect condition. Wear, scratches, or a deformity always cause a reduction in pump efficiency.

Now, using a jet of compressed air, check that the water ducts in the power head are clear.

Remember to blow a jet of air through the tell-tale hole. Because of its small diameter, it is easily clogged with debris.

> **! IMPORTANT**
> **Despite their high cost, never hesitate to change parts that do not appear to be in perfect condition. The life of your engine is at stake. Most manufacturers provide kits that include the impeller, wear plate, casing and seals.**

Reassembly

- Replace the wear plate with a new gasket, then fit the impeller, remembering to replace the key.
- Wet the impeller with soapy water.
- Refit the pump casing, including its rubber gasket, while turning the drive shaft clockwise to position the impeller blades correctly.
- Screw on the pump casing.
- Remount the lower leg.

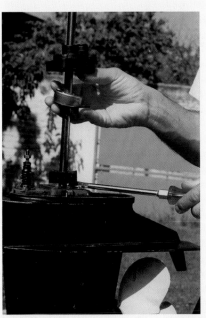

1 To enable you to work on the water pump more easily, it is best to clamp the lower leg in a vise, protecting it so that the surface isn't damaged.

2 Unscrew the pump casing, then pull off the casing, impeller and wear plate.
   Caution: remember to pick up the locking key.

3 Scrape off any deposits, clean with petrol, then blow or wipe the drive shaft dry.

4 Check the water pump parts for scratches on the casing and wear plate, deformities, wear on the impeller and the condition of the key. If in doubt, change any defective parts.

5 Note the position of the key. It often tends to slip out of place when you put the impeller back on.

6 When replacing the water pump casing, make sure the impeller blades are oriented correctly. To simplify this job, wet the impeller with soapy water and turn the drive shaft clockwise.

7 Be sure to clean the water intake grids.

Remove, check and replace the propeller

Simple

● 20 minutes

● Basic tool kit, cotter/split pin, new shear pin for small engines, marine grease

● Remove the propeller

Remove the cotter/split pin. Jam a piece of wood between the propeller blade and the anti-cavitation plate. Unscrew the propeller nut or end cap. Remove the propeller. If needed, tap behind the propeller with a mallet. Remove the thrust washer.

> **!** **WARNING**
> **To avoid an accidental start during this procedure, remove the kill cord or the spark plug lead.**

● Inspection and maintenance

◆ Check that the propeller blades are not worn, damaged, warped or corroded. It is sometimes possible to straighten a bent blade. Work on a workbench. Adjust the blade using a mallet.

◆ Check that the edges of the blade are not too deteriorated. If they are, file them to a uniform thickness.

◆ Replace the propeller if one or more of the blades are worn, damaged or cracked. Damaged blades cause excessive and destructive vibration.

> **High engine revs and slow boat speed are signs that you should check the propeller's rubber bush.**

Check that the shear pin (2–4hp engines) or the splines are not damaged. Clean the grooves in the propeller and the propeller shaft. Grease the shaft.

● Remount the propeller

◆ Install the propeller taking care to remember the thrust washer.

◆ Tighten the nut or propeller end cap until the slots are aligned with the hole in the propeller shaft.

◆ Insert a new cotter pin and bend the ends.

◆ Make sure that the engine is in neutral and spin the propeller by hand. It should turn freely.

> **!** **IMPORTANT**
> **Maintain your propeller in good condition to ensure the best performance for your boat.**

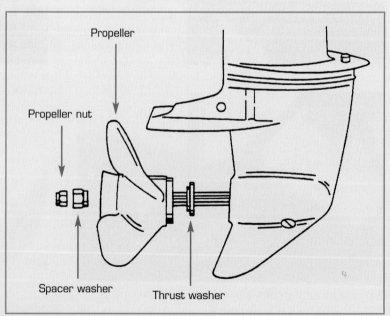

Propeller

Propeller nut

Spacer washer

Thrust washer

1 Removal: remove the split pin or tab washer.

2 Put a block of wood between the anti-cavitation plate and the propeller blades to unlock the nut.

3 Tap the hub with a mallet, if necessary, to loosen it.

4 After cleaning and greasing the shaft, remount the propeller, screw the nut on, then install the split pin or bend the washer tab.

Checking the compression

Technical

15–30 minutes, depending on the number of cylinders

Compression gauge, spark plug spanner

The compression is checked to find out if there is a leak in the head gasket or the cylinder rings. It must be done with a compression gauge.

The compression gauge allows you to check whether the compression is the same as that indicated by the manufacturer.

How to do it

- Remove the spark plugs, making sure to note the location of each lead.
- Push the compression gauge into the spark plug socket (for gauges with a rubber end) or screw in the type that has a thread.
- Remove the kill switch lanyard or push the kill switch to protect the electronic block by short circuiting the ignition system.

Pull on the starter cord to spin the engine until the gauge needle stabilises, or turn on the starter for five seconds.

> **! IMPORTANT**
> During this reading the throttle must be wide open, ie put the throttle in the 'fast' position.

1 Screw the compression gauge into the spark plug socket.

2 When you pull on the starter cord:
- open the throttle completely;
- push the stop button.

Checking the compression gauge reading

Push the reset button on the gauge to reset it to zero. Repeat the operation on the other cylinders.

What the readings mean

- The readings for each cylinder should be at least equal to the volumetric ratio number.
- The minimum compression should be 7 bars (101lb/sq in).
- Low compression in all of the cylinders indicates worn piston rings.

- Variation between the cylinders should not be greater than 1 bar (15lb/sq in). The cylinder with the lowest reading is defective.
- On a four-stroke engine, lower compression in one cylinder could indicate a burnt valve, or on a two-stroke engine, a defect in the piston ring.

> **! IMPORTANT**
> **Systematically check the compression before tuning the engine as it is impossible to tune an engine with weak or irregular compression. It is therefore essential to correct any compression problems *before* tuning the engine.**

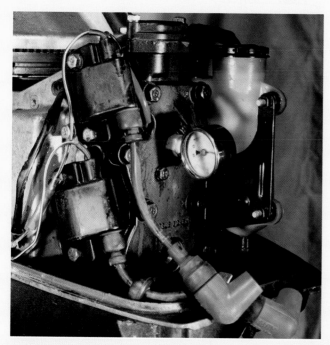

3 Two or three pulls on the starter cord should be enough to stabilise the compression gauge needle.

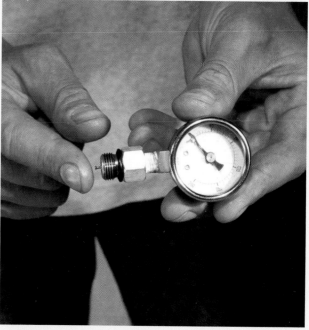

4 Remember to reset the needle before testing another cylinder.

Preparing the fuel mix

Simple

● 5 minutes

● Measuring cup, oil that meets National Marine Manufacturers Association (NMMA) grade specifications, petrol

A proper fuel mix is essential for all outboard engines under 5hp and those without an automatic oil/fuel mixing system. Fastidious fuel mix preparation is an unavoidable task for every boat owner. It involves achieving the manufacturer's recommended oil/fuel mix – do it methodically.

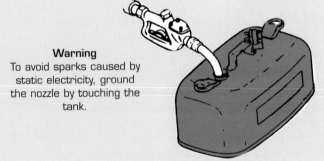

Warning
To avoid sparks caused by static electricity, ground the nozzle by touching the tank.

! WARNINGS AND ▪ PRECAUTIONS

◆ Stop your engine while refuelling.
◆ Petrol is very flammable and explosive. When you mix it with oil, or during refuelling, do not smoke and make sure you are away from open flames or sparks.
◆ When you are refuelling, avoid electro-static sparks by keeping the fuel nozzle in contact with the tank. For the same reason, do not use a plastic funnel.
◆ Never overfill the tank.
◆ After filling the tank, put the fuel tank cap back on.
◆ Prepare the fuel mixture outdoors or in a well-ventilated place.
◆ Follow the manufacturer's recommendation for the quality and type of oil.
◆ Use only marine grade oil that meets **NNMA/BIA TCW** specifications.

MAKE YOUR MIX

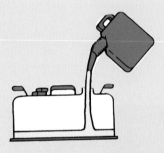

Pour a small quantity of petrol into the can.

Now pour in the amount of oil needed for the entire mix. A measuring cup, available at most dealers, greatly simplifies the operation.

● How to do it

◆ Pour a small quantity of fuel in the tank.
◆ Add the amount of oil required for the entire mix into the tank.
◆ Add the petrol.

Top up with petrol to complete the mixture.

AMOUNTS AND PERCENTAGES		
	Oil	Petrol
Running-in period	1 litre	25 litres
After running in	1 litre	100 litres (new style engine)* 25–50 litres (older style engine)

*Check your engine manual

Cleaning the fuel filter

Simple

- 10 minutes
- Cleaning materials

To protect the fuel pump and carburettor from impurities it is important to clean the fuel filter.

This should be done every 100 hours and before each winterisation.

● How to do it

- Find the fuel filter on the engine.
- Unscrew the bowl.
- Remove the filter.
- Examine the filter.
- Make sure there are no cracks or signs of deterioration.

- Clean the bowl and the filter in a container with petrol and a soft brush (a toothbrush does the job well).
- If possible, dry with compressed air.
- Replace the filter in the bowl.
- Screw in the bowl.

1 Check the condition of your filter regularly.

2 On most engines, the fuel filter is accessible. You just unscrew the bowl to get to the filter.

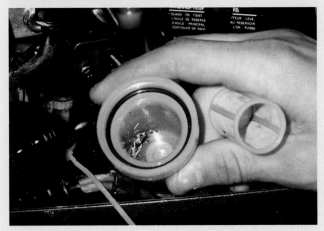

3 A mess of debris!

4 Clean the filter and the bowl with a brush and petrol. If possible, finish the cleaning by drying with compressed air.

Cleaning the carburettor

Complex

- 30 minutes to 2 hours, depending on the number of carburettors and their ease of access
- Basic tool kit, tape, petrol, small brush, compressed air, engine or carburettor cleaner

Even though the fuel system has a filter, some debris always ends up accumulating in the bottom of the bowl. Also, the oil mixed in the petrol tends to settle in the bottom of the carburettor passages. If the debris is excessive, it can affect the carburettor function. Regular cleaning of the carburettor is recommended.

Although the bowl can be cleaned without removing the carburettor, for a thorough clean you will have to remove it.

● Removing the carburettor

Removal and reinstallation of the carburettor will vary depending on the type of engine, but the process is always the same.

Disconnect all the fittings (fuel lines, choke, filter, throttle, etc) before removing the carburettor itself.

Before starting to dismantle the float bowl and the different injectors, it is essential to clean the *exterior* of the carburettor to prevent any grime and grit from contaminating it.

● Cleaning

Be methodical. The carburettor is a delicately balanced, complex part. The simplest way to clean it is by using petrol and a brush.

1 Carburettor removal
Don't remove the carburettor screws until you have disconnected the fittings, ie the choke, throttle and fuel lines.

2 Cleaning
Cleaning is done with clean petrol and a brush. If the carburettor is very dirty, use the type of carburettor or engine cleaner that you would use for winterising. Soak the carburettor to remove the stubborn deposits.

Fill a container with clean petrol. Place the carburettor in it.

The petrol will initially dissolve greasy residues and salt, then you can finish off with the brush.

● Removal of the float bowl and the jets

◆ Turn the carburettor upside down and loosen the screws on the float bowl.
◆ Remove the bowl and the float. *Warning*: different manufacturers attach these in different ways, depending on the engine type or size. Generally, you can slide the float pin out sideways.
◆ Remove the needle valve. Unscrew and remove the high-speed needle. Unscrew and remove the idle jet.

 The two parts of the idle assembly are located on the carburettor exterior and are accessible without removing the carburettor.

● Inspection and cleaning

Carefully examine the bowl because this is where deposits tend to accumulate.

There are three main types of deposits:

◆ A very fine brown deposit made up of minute particles of rust from oxidation of the fuel tank connector or your fuel tank.
◆ A sticky, oily deposit from the fuel mixture.
◆ A whitish deposit that is formed when the engine is shut off and stored in a damp place.

If the deposits are light, a simple clean up with petrol and a brush is sufficient. If they are heavy, oily deposits, you need to use a special aerosol carburettor cleaner to dissolve the residue present in the bowl and also in all the carburettor channels.

Rinse with clean petrol. Blow into all the holes and jets.

Compressed air is ideal for cleaning a carburettor.

 IMPORTANT
Do not use any metallic object to unclog the jets or channels.

3 Disassembly
Loosen the screws holding the bowl, then remove it. If the bowl is stuck, tap it lightly with a mallet.

4 Remove needle valve
Slide out the pin holding the float with needle-nosed pliers, then remove the needle valve.

The needle valve tends to wear out at its seat. Check that it doesn't show any trace of wear and that it slides freely in its shaft. Otherwise, replace the needle valve (see page 97: Adjust the float level).

5 Remove the main jet. Select your screwdriver size carefully so as not to damage the jet during removal.

6 Clean with clean petrol, then blow into all the holes.

7 For carburettors equipped with a throttle slide, unscrew the top of the carburettor then remove the throttle slide.

8 To remove the throttle slide, compress the spring.

● Check the float level

At the same time, check the float level. To do this, turn the carburettor upside down; normally the float should be parallel to the seal base. The best thing, of course, is to check the float level by measuring it and making sure it is the same as that specified by the manufacturer.

If needed, adjust the level by slightly bending the float tab.

● Reinstall the carburettor

Reassemble following the inverse order of disassembly. However, before remounting the bowl, check the position of the seal. Before reinstalling the carburettor, adjust the high-speed needle. Screw it in all the way without locking it, then unscrew one and a half turns (basic adjustment).

9 Check the throttle slide/needle assembly. It should slide easily within the carburettor body. Remove the throttle slide/needle assembly.

10 Check the float level with a measuring tape.

11 Adjust the float level by slightly bending the float tab.

12 The carburettor is clean; the float level is set; you can reinstall your carburettor.

Adjusting the idling rate

Simple
- 5 minutes
- Screwdriver

Engine type
2 and 3hp engines with a throttle slide carburettor.

Adjust the engine idle in a test tank, or on the boat in the water. The adjustment has to be made when the engine is warm.
- Find the idle adjustment screw.
- Start the engine.
- Let the engine warm up until it reaches its operating temperature.
- With the engine idling, adjust the throttle slide adjustment screw until you get a stable idle speed.
- When you tighten the screw, the idle speed increases; conversely, when you loosen it, it decreases.
- Allow 15 seconds between each adjustment.

Screw in = revolutions/minute (rpms) increase

Unscrew = revolutions/minute (rpms) decrease

Once you have found the throttle slide adjustment screw, adjusting it is simple. When you unscrew the screw, the speed decreases and when you screw it in, the speed increases.

Adjusting the idling rate and fuel/air mix

Technical

- 5 minutes
- Basic tool kit

Engine type

All engines with a **throttle valve carburettor**.

These adjustments are made to adjust the engine speed and fuel/air ratio at idle.

● Engine tests and adjustments

The adjustments should be made with the engine warm.

Test the engine and adjustments in a test tank or on the water. Locate:

◆ The idle mixture needle
◆ The idle jet
◆ The adjustment screw for the throttle valve

Adjusting the low speed needle

Unscrew = more rpm

Screw in = less rpm

Adjusting the screw alters the idle mixture. When you unscrew the idle mixture screw, the mix is richer. Screw it in and the mix is leaner. Be careful not to lock the screw into its seat.

Look for the manufacturer's data regarding the recommended idle mixture adjustment. If you don't have it, follow the procedure below:

◆ Start the engine.
◆ Let it warm up (at least 3 minutes under load).
◆ With the engine at idle, slowly unscrew the idle screw until the engine revs decrease.
◆ Slowly screw in the idle screw, counting the fractions of turns until the motor runs regularly and the revs go up.
◆ Continue to screw in the screw until the engine starts to miss.
◆ Then unscrew half the fractions of turns counted previously. Then adjust the idle by turning the throttle valve adjustment screw.

Adjusting the throttle valve screw

Unscrew = more rpm

Screw in = less rpm

Unscrew the throttle valve screw and the revs decrease. Screw it in and they increase.

Adjusting the fuel/air mix

Complex
- 15 to 30 minutes
- Basic tool kit, spark plug spanner

Engine type
2 and 3hp engines with a throttle slide carburettor.

Throttle slide carburettors have a conical needle that adjusts the flow of fuel according to the opening of the throttle slide.
   The needle is normally clipped on the third notch. Clipping it higher makes the mix leaner; lowering it makes it richer.

● Engine tests and adjustments

Testing should be done with the engine warm.

◆ Warm up the engine until it reaches its operating temperature.
◆ Test the engine by running it in gear with the throttle fully open (in the 'fast' position) for three minutes.
◆ Stop the engine. Remove the spark plug. If the mix is well adjusted, the spark plug will not be moist or overheated.
   If during the test the revs drop or don't increase after opening the throttle slide two thirds of the way, the mixture is too rich. Move the clip to the upper notch.
◆ Repeat the test. Keep the throttle fully open. The engine must reach maximum high revs when the throttle slide is fully open.
◆ Stop the engine. Remove the spark plug. If the central electrode and insulator look whitish, the mixture is too lean. If the central electrode and insulator are moist and slightly oily, the mixture is still too rich. You have to move the clip up another notch on the needle.

> ✸ You must check the fuel level in the bowl before continuing the adjustment.

Once the needle is at the correct height, readjust the idle speed.

> **!** **IMPORTANT**
> **A light to mid brown colour indicates good combustion.**

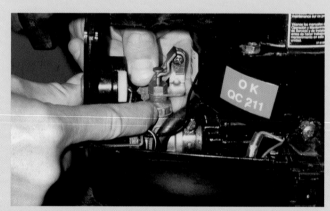

1 Unscrew the throttle slide support.

2 Remove the throttle slide.

3 When replacing it, take care to centre it in its sprocket.

◆ Points to note

◆ If the engine misses while accelerating, it is probably because the mix is too lean. Unscrew the idle mixture adjustment screw until the engine accelerates without missing.

◆ If the engine routinely stalls at idle, it is possible that the idle jet is blocked. Remove the idle jet and the idle mixture adjustment screw and blow the jets and carburettor apertures with compressed air. Replace the jet and idle mixture adjustment screw. Test again. If the problem persists, a deeper cleaning and check of the carburettor are needed.

◆ If the idle is unstable there is definitely an air leak. It could be located between the carburettor, the valve box and the engine. Consult your service engineer.

Setting the needle height

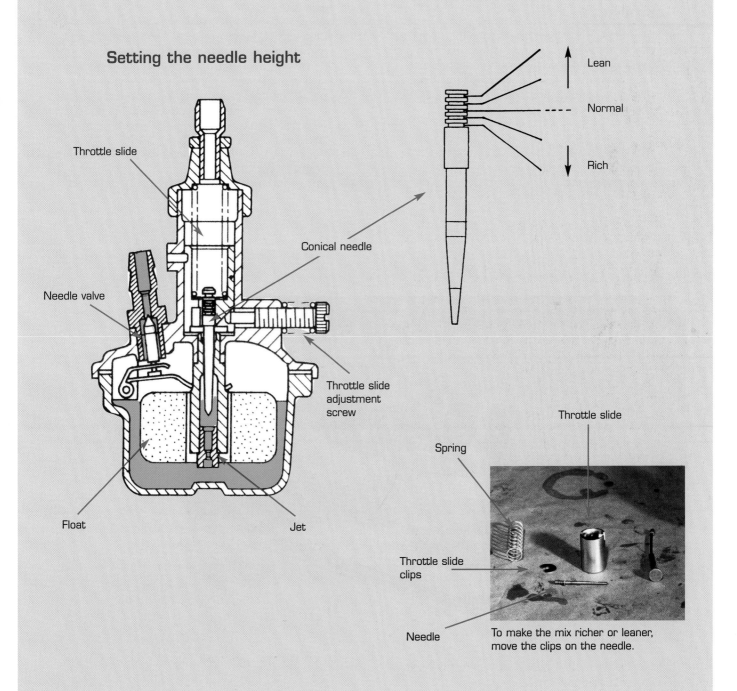

Throttle slide

Conical needle

Needle valve

Throttle slide adjustment screw

Float

Jet

Lean

Normal

Rich

Spring

Throttle slide

Throttle slide clips

Needle

To make the mix richer or leaner, move the clips on the needle.

Adjusting the bowl fuel level

Complex

- 5 to 10 minutes once the carburettor is removed
- Screwdriver, measuring tape, small flat pliers

● Throttle slide and throttle valve carburettors

The level of fuel in the bowl determines the performance of your engine. A high level causes too rich a mix and, conversely, a low level causes too lean a mix.

The fuel level in the bowl is controlled by a float connected to a needle valve that closes the fuel inlet when the correct level in the bowl is reached.

● How to do it

- Remove the carburettor.
- Remove the bowl.

- Remove the float, generally by sliding its pin sideways.
- Remove the needle valve.
- Check the bowl level height. The exact measurement is in the manufacturer's repair manual. If you don't have this information, turn the carburettor upside down with the needle valve and float in place, then check that the float is exactly parallel to the base of the carburettor bowl. If needed, the height can be adjusted by slightly bending the float bowl tab.
- Reinstall the bowl, taking care to note the position of the seal.
- Reinstall the carburettor.

1 After cleaning the carburettor, you can remove the bowl. Work on a flat, clean, well lit surface.

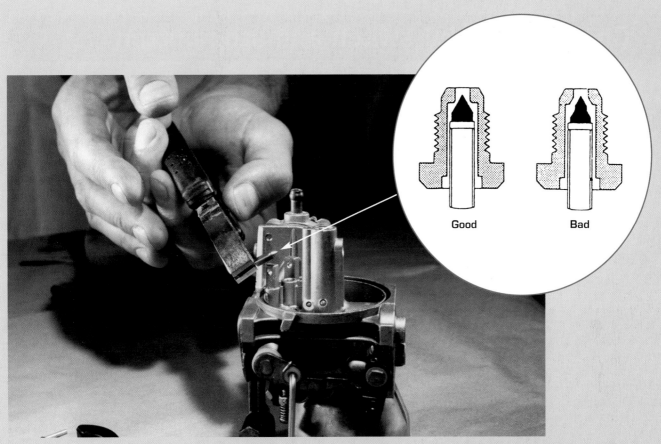

2 The bowl and float have been removed. Remove and check the needle valve. It should slide freely in its shaft and should not show any signs of wear.

3 Measure the height with the carburettor upside down.

4 You can adjust the bowl level by bending the float bowl tab. Beware: this job is very delicate. Make small adjustments.

Complex

- 30 minutes
- Multimeter, small tools

Your engine refuses to stop or, conversely, it refuses to start. First, check the kill switch. This is usually connected to the stop button.

● How the system works

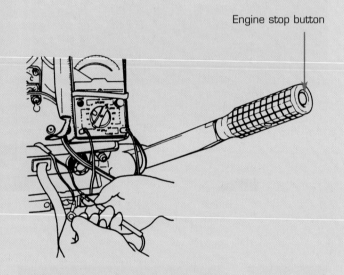

Engine stop button

When the kill cord is removed, the engine stops (no ignition).

When the cord is in place on the kill switch, the system is on.

To stop the engine when the kill cord is attached, push the stop button until the engine stops.

If you don't push the stop button, the engine will stay on.

● System check

You will need a multimeter to do this.

- ◆ Disconnect the kill switch: two wires, the first connected to ground, the second to the electronic box.
- ◆ Adjust your multimeter to ohmmeter (scale 1).
- ◆ Connect the clip and cord to the kill switch.
- ◆ Connect the multimeter to each wire.
- ◆ The current shouldn't flow; the resistance is infinite.
- ◆ Briefly push the stop button.
- ◆ The current should flow; the ohmmeter should show a weak reading.
- ◆ Remove the cord. The current should flow; the ohmmeter should show a weak reading.

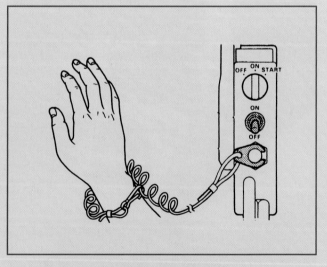

Check the kill switch
Regularly check the operation of your kill switch. To check the circuit, disconnect the kill switch and stop-button wiring.

This test can be done without disconnecting the earth wire.
To do this, connect the multimeter clip to earth on the engine.

To avoid a sudden loss of control, replace the kill switch or stop button if there is the slightest doubt as to their proper functioning (strong resistance, defective waterproofing, no contact or intermittent contact).

Regularly check the proper functioning of your kill switch.

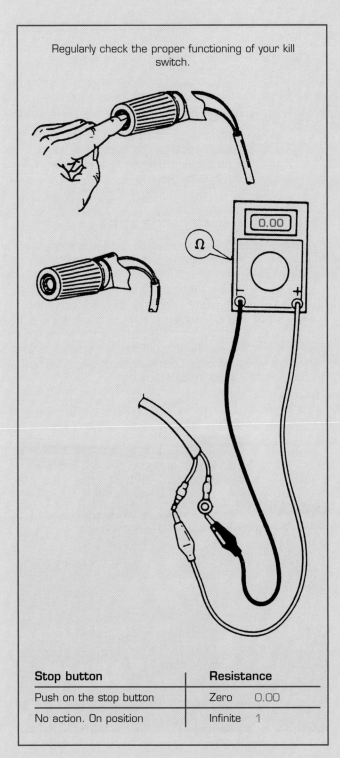

Stop button	Resistance	
Push on the stop button	Zero	0.00
No action. On position	Infinite	1

To check the circuit, disconnect the kill switch and stop button wiring.

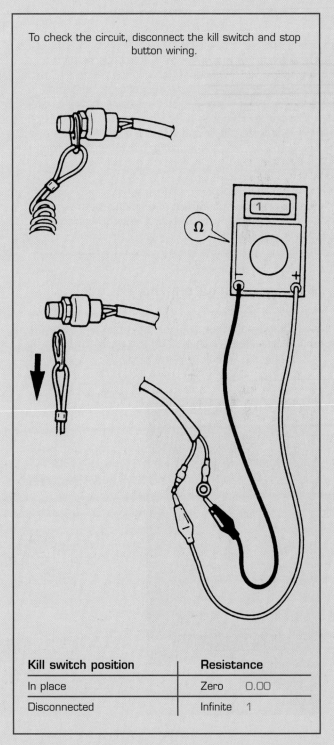

Kill switch position	Resistance	
In place	Zero	0.00
Disconnected	Infinite	1

Checking the battery

Technical

- 15 minutes
- Basic tool kit

The battery is a reservoir of electrical energy. As such, it supports the many electrical functions needed for the boat's performance. It is normally recharged by an alternator. However, during high energy consumption or during winter storage, the battery potential drops.

The battery charge has to be checked every 100 hours or once per season. There is a difference between checking the electrolyte level and checking the charge state.

> ✷ High water consumption may be a sign of battery overcharging caused by a faulty regulator.

● Checking the electrolyte level

Traditional batteries need to have their electrolyte levels checked. Each element has to be permanently submerged to a depth of 1 to 1.5cm. Most batteries have a level indicator to show you this.

Top up with distilled or demineralised water as needed using a small funnel.

> **!** IMPORTANT
> **When adding water during the winter, immediately recharge the battery to avoid the risk of freezing caused by the water and acid not fully mixed.**

● Checking the charge state

The usual way to check the battery's charge state is to check the electrolyte density with a battery hydrometer.

Checking one at a time, take just enough fluid from each cell with the hydrometer until the float counterweight floats freely. Make sure you don't spill any drops on the boat or on your clothing because the sulphuric acid is extremely corrosive. Hold the hydrometer vertical at eye level and check the specific gravity reading to establish the electrolyte density.

Undercharged

Half charged

Charged

A battery hydrometer allows you to determine the battery charge state. Measure the specific gravity in each cell. Note that the readings are affected by temperature.

If the electrolyte level is low, add only distilled or demineralised water.

A density of 1.26 indicates that the battery is totally charged. A density of 1.20 indicates that the battery is charged at 50% of its capacity.

If the density doesn't exceed 1.10, the battery is 80% discharged. It must be recharged.

Many hydrometers consist of nothing more than coloured zones on the float to indicate the battery charge state. For a more precise reading, the specialists use a compensation thermometer that indicates the correction to apply if the battery temperature is below or above 15°C.

> **❗ IMPORTANT**
> **Never measure the electrolyte density immediately after having topped up the battery.**

● Maintenance-free batteries

Maintenance-free batteries don't allow access to the electrolyte.

A voltmeter measured in tenths of a volt is needed. Check your battery when cold, ie when it has been at rest for at least two hours.

A reading of 13 volts shows that the battery has all of its potential. At 12.2 volts it is 50% charged, and at 11.9 volts it is at 20% capacity.

Certain batteries (Freedom, Vetus) are equipped with a charge indicator window, allowing you to check its condition at a glance. If the window is green, the battery is charged. If it is black, charge the battery until the green dot reappears.

A yellow or light dot indicates that the battery is malfunctioning.

> **❗ IMPORTANT**
> ◆ **When handling a battery, make sure you do not put metal tools across its terminals.**
> ◆ **Never interchange the battery terminals.**
> ◆ **There is a + or – on each terminal. To avoid confusion, the diameter of the positive terminal is larger than the negative terminal. The positive terminal feeds the different electrical devices. The negative terminal is earthed.**
> ◆ **Don't smoke near a battery while it is charging or has recently been charged.**
> ◆ **If the battery acid comes in contact with skin or eyes, wash the skin immediately with a mild soap. Rinse the eyes in fresh water immediately and see a doctor.**

For a maintenance-free battery, use a voltmeter measured in tenths of a volt or a numeric voltmeter.

Charge state	Voltage
100%	13.00 v
75%	12.60 v
65%	12.45 v
50%	12.20 v
25%	12.00 v

A glance at the charge indicator window tells you its condition.

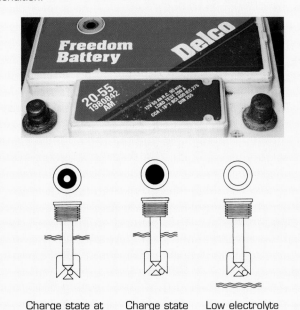

Charge state at 65% or +   Charge state below 65%   Low electrolyte level

Recharging the battery

Technical

- Time taken depends on the initial charge state of the battery
- Battery charger

Apart from the density or voltage reading, we know that a battery is insufficiently charged when the starter motor has trouble cranking or doesn't turn at all when you're trying to start the engine. The battery must be recharged with an electric battery charger.

Remove the battery plugs or caps

◆ Check the fluid levels.
◆ Connect the charger's + terminal (red wire) to the battery's + terminal and the charger's – terminal (black wire) to the battery's – terminal.
◆ Before turning the charger on, make sure that the voltage setting on the charger matches the battery voltage. This will generally be 12 volts. Select the charge rate. This depends on the battery capacity. Calculate 2 amperes for every 10 amperes of capacity. Don't set the charge rate higher than that or you may risk damaging the battery.
◆ You can tell if the battery is fully charged when the liquid in every cell bubbles.
◆ Turn off the charger.
◆ Disconnect the charger from the battery.

> ! **IMPORTANT**
> **To avoid the risk of explosion always turn off the charger before disconnecting the battery.**

If, at the end of a long recharge, the electrolyte density in each cell doesn't exceed at least 1.20, the battery is not in an optimal state. If one of the elements doesn't want to take a charge and has a density that is still less than 1.10, change the battery.

Chronic battery discharge is often due to age. A battery's life expectancy is about four years.

> ✳ **Note**
> **The recharged battery will yield about 75%: a 40A/h battery will take about 52A/h to charge.**

1 Clean the battery terminals.

2 A little petroleum jelly applied to the battery terminals reduces corrosion.

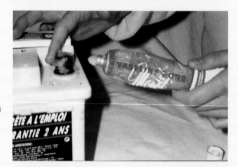

3 Remember to remove the battery plugs/caps before recharging the battery.

4 Do not exceed the charge rate. Example: a 45A/h battery = 9A maximum allowed.

Checking the oil pressure

Complex

- ◆ From 15 to 30 minutes
- ◆ Basic tool kit, oil pressure gauge

All four-stroke outboard engines are equipped with bearing surfaces lubricated by pressurised oil (crankshaft/camshaft) and an oil pressure light connected to an audible alarm.

The pressure of the oil within the circuit affects the engine's longevity.

> **! WARNING**
> **When the oil pressure light comes on or the alarm sounds, the engine is already malfunctioning due to lack of oil pressure.**

First check the oil level. If it is correct, checking the oil pressure with an oil pressure gauge will give you a reading of the pressure in the circuit.

This reading will indicate the degree of engine wear.

● Check

Make sure you have an oil pressure gauge that fits into the oil pressure sending unit. This task is done while the engine is cold and until it reaches running temperature.

Remove the oil pressure sending unit.
Screw the oil pressure gauge in its place.
Start the engine.

● What the test tells you

When the engine is cold, the oil pressure is always higher than the manufacturer's specification* because the oil's viscosity is always higher when cold. When the engine is warm, it should meet the manufacturer's specification.

A low reading at idle is a sign of a significantly worn out oil pump assembly and/or oil seals.

*Average pressure: 3.5 to 4.5 bars (51 to 65 lbs/sq in.)

- ◆ If the pressure is 1 or 2 bars under the manufacturer's specification, the engine is starting to wear out.
- ◆ Pressure of more than 2 bars under means that there is a real problem in the lubrication system.
- ◆ If the pressure is 1 bar under, it is worth checking the problem straight away to avoid serious engine damage.

1 Find the oil pressure sending unit.

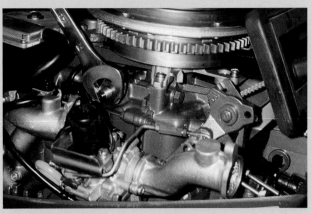

2 Screw in the oil pressure gauge.

3 The reading should be between 3.5 and 4.5 bars.

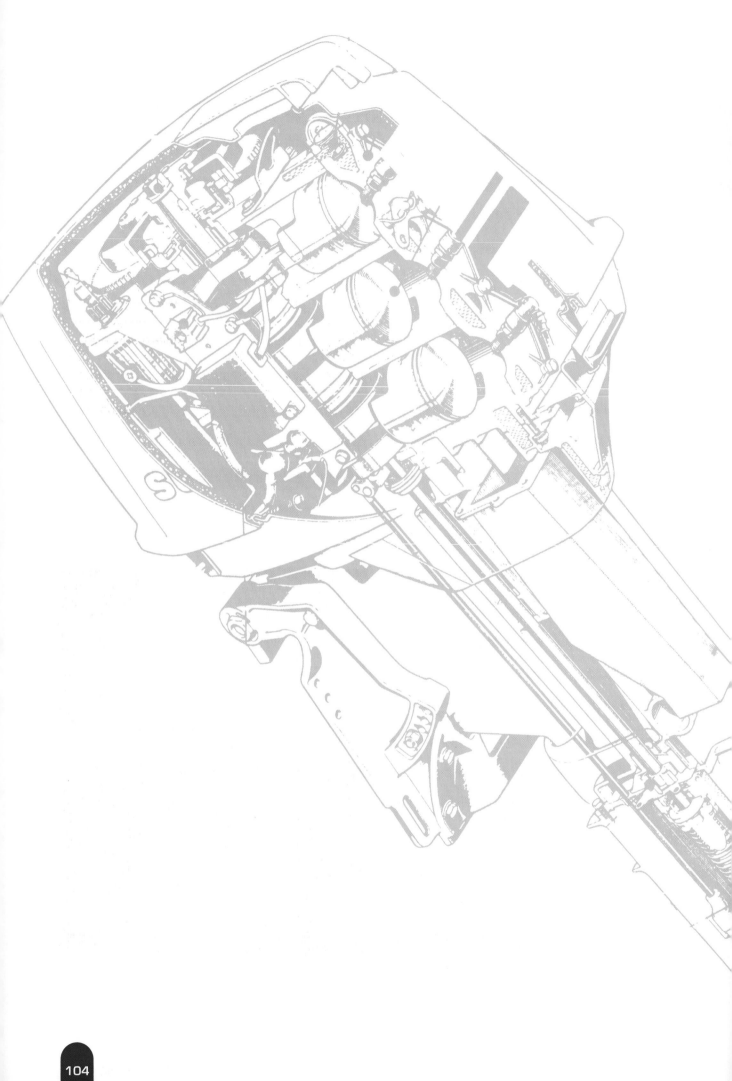

BREAKDOWNS

HOW DO YOU IDENTIFY THE PROBLEM? Hunt down the first signs of wear, any unusual noise, or vibration, or leaks before the breakdown happens. Make checks to locate the cause of the problem. An engine rarely breaks down without a warning sign. The traditional weaknesses of our engines have known symptoms. Learn how to recognise them. Also learn how to tell which symptoms are benign and which are not.

The following tables will help you find the cause of the breakdown more easily.

BREAKDOWN TABLES

When confronted with a breakdown or a problem, find the description below that corresponds to it. Read the corresponding numbers. Each number refers to a possible cause of the breakdown and its solution.

> **! IMPORTANT**
>
> Before starting any work on the engine you must identify the cause of the malfunction. For example, the engine refuses to start. Two causes are possible: a faulty electrical circuit or a fault in the fuel system.
>
> By simply checking the spark plugs it is often possible to identify the cause of the breakdown.
>
> ◆ Dry spark plugs indicate a faulty fuel supply.
> ◆ If there is no spark at the plug, you have to look for a fault in the ignition system.

Problem description	Number to check
The engine has trouble starting.	1, 3, 4, 6, 9, 11, 12, 19, 23, 29, 33, 34, 38, 39
The engine refuses to start.	1, 2, 3, 6, 7, 8, 9, 10, 22, 29, 31, 33, 35, 36, 37, 38, 39, 41, 42, 62
The engine starts but quickly stalls.	38, 39, 41, 42, 62
The engine doesn't idle.	1, 4, 14, 15, 18, 23, 27, 37
The engine coughs.	7, 8, 3, 9, 10, 14, 32, 33, 34
The engine has no power.	7, 8, 9, 10, 3, 1, 4, 17, 20, 21, 22, 23, 24, 25, 26, 27, 29, 39, 49
The engine overheats.	57, 58, 59, 60, 61, 5, 39
The engine runs too fast.	49, 50, 48, 51
Excessive fuel consumption.	27, 29, 39, 18, 16, 11, 49, 51, 52, 54
The engine vibrates.	1, 29, 28
The gears won't engage.	45, 46, 47
Poor boatspeed.	48, 50, 51, 52, 53, 54, 55

Probable causes		Solutions
1	The fuel is old or contaminated.	Drain the fuel tank and fuel lines. Fill with new fuel.
2	The tank is empty.	Fill the tank.
3	There is water in the fuel.	Drain the tank and fuel lines.
4	The mixture has too much oil.	Drain the tank and fuel lines.
5	The mixture has too little oil.	Drain the tank and fuel lines. Refill the tank according to the oil/petrol mix specification.
6	The starter doesn't work.	Check the starter controls.
7	The fuel line is badly connected.	Reconnect the line and check all the connection fittings.
8	The fuel line is crimped.	Uncrimp line and make sure it isn't pierced.
9	The fuel filter is clogged.	Dismantle the filter and clean it.
10	No air inlet at the fuel tank.	Open the air inlet located on the fuel tank cap.
11	The fuel pump diaphragm is defective.	Check the fuel pump diaphragm.
12	The needle valve is stuck.	Remove and clean the carburettor; check that the needle valve slides easily.
13	Sediment in the carburettor.	Remove and clean the carburettor.
14	The fuel mix screw is out of adjustment.	Basic adjustment: without forcing, screw in the fuel mix screw, then unscrew it one and a half turns.
15	The idle screw is out of adjustment.	Adjust the idle (with the engine in gear).
16	The bowl level is incorrect.	Remove the carburettor, check and if necessary adjust the bowl level.
17	Poor carburettor synchronisation.	Consult a specialist.
18	The fuel mix is too rich in petrol.	Check the fuel pump (pierced diaphragm). Check the bowl level.
19	The fuel mix is too lean in petrol.	Remove and clean the carburettor; check the bowl level.
20	The piston and cylinder are scratched.	See a specialist.
21	The piston rings are worn.	See a specialist.
22	Poor fuel supply to the carburettor.	Check the supply of fuel to the carburettor by disconnecting the carburettor fuel line. Pull the starter cord or turn on the starter.

Probable causes		Solutions
23	Weak compression.	Check the compression and compare it to the manufacturer's specifications.
24	A leak at the engine head or casing.	See a specialist.
25	The crankshaft seals are leaking.	See a specialist.
26	The inlet valves are damaged or deformed.	Remove the valve box, check and adjust if necessary.
27	The engine has heavy carbon deposits.	Run the engine with a proprietory engine cleaning product.
28	The piston rod bearings are defective.	See a specialist.
29	The spark plugs are dirty.	Remove and replace the spark plugs.
30	The on/off switch is disconnected.	Reconnect it.
31	The engine is in gear.	Put the gear shift in neutral.
32	The spark plug type is wrong.	Remove, replace the spark plugs. Follow the manufacturer's specification.
33	Bad connection.	Check and retighten the connections.
34	Bad earth.	Check the earth wires and re-establish the connections.
35	Excessive humidity in the electrical circuit.	Dry, then spray with a water-repelling fluid/penetrating oil.
36	The stop button or the kill switch is defective.	Replace if defective.
37	The coil is defective.	See a specialist.
38	The electronic ignition system is defective.	See a specialist.
39	The timing is out of adjustment.	See a specialist.
40	The fuse is defective.	Replace the fuse and check the electrical circuit.
41	The battery is flat.	Charge the battery.
42	The battery terminals are loose or corroded.	Clean, then re-tighten the terminals.
43	The starter solenoid is defective.	Check and replace the solenoid if necessary.
44	The starter motor is jammed.	Free the gear and spray with penetrating oil.
45	The gear shift controls are out of adjustment.	Adjust the controls.
46	The gear selector is broken.	See a specialist.

Probable causes		Solutions
47	The gears are worn or broken.	See a specialist.
48	The shear pin is broken.	Check the propeller and change the pin.
49	The propeller blades are damaged, deformed, or worn.	Replace the propeller.
50	The engine position is incorrect.	Reposition the engine on the boat. Adjust the trim.
51	Wrong propeller.	Change the propeller. The correct propeller is the one that will take the full load of the engine at full throttle.
52	The rubber propeller bush slips.	See a specialist.
53	The boat is overloaded.	Observe the recommended maximum load.
54	The hull is fouled.	Clean the hull.
55	The hull is deformed.	See a specialist.
56	The load is unbalanced.	Distribute the load evenly.
57	The water inlet is clogged.	Clean the water inlets.
58	The water pump impeller is worn.	Remove the lower leg, replace the impeller.
59	The water pump case and wear plate are scratched.	Remove the lower leg and check the water pump.
60	The thermostat is stuck closed.	Remove the thermostat, check its opening temperature and replace if needed.
61	The water circuit is clogged.	See a specialist.
62	Fuel cock off.	Open.

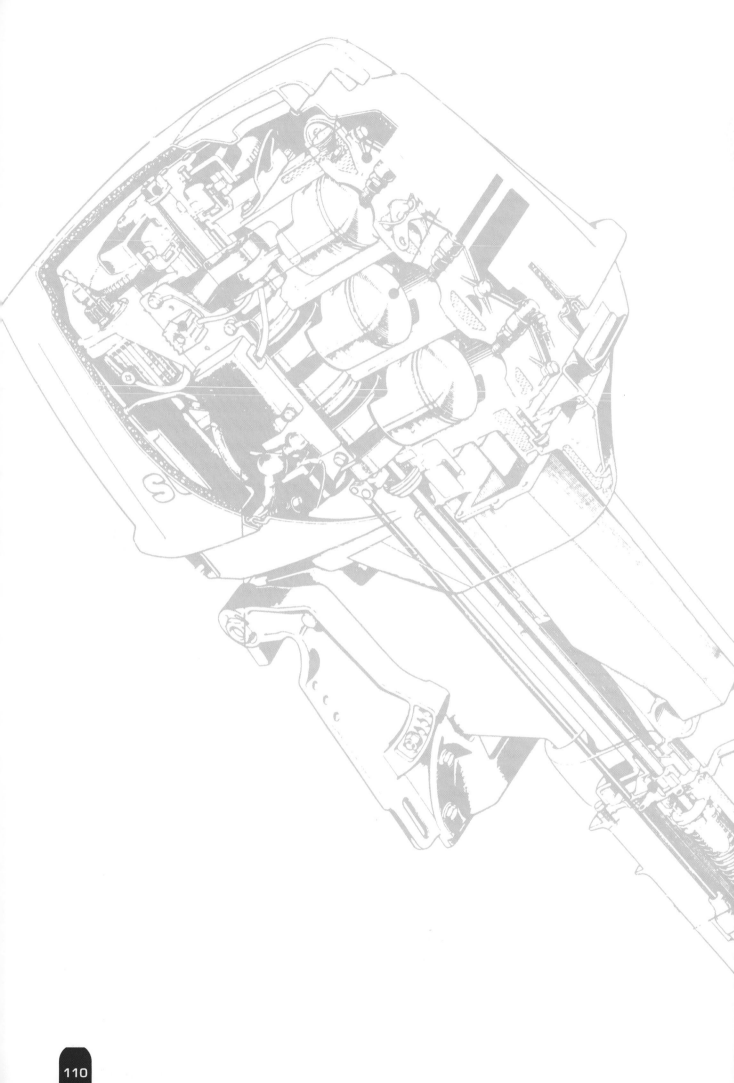

WINTERISING

IT IS NEARLY THE END of your boating season. It is time to think about winterising your engine to protect it from corrosion. This is within everyone's capability, especially on the smaller and medium sized engines.

However, if you feel that this procedure exceeds your level of competence, don't hesitate to hand it over to a professional as soon as possible because corrosion acts with surprising speed.

Recipe for good winterisation
Good tools, appropriate products; a little time and care.

Freshwater inlet

Towards the engine

Towards the engine

A simple solution: just connect a garden hose to the easily accessible fitting to flush the engine without having to put it in gear (Yamaha flushing system).

Whatever the make of your engine, the winterisation procedure is more or less the same.

Desalting and lubrication

If your engine is used in salt water, you must run it in fresh water to dissolve the salt deposits in the cooling system. This must be done for at least 30 minutes. There are a number of easy ways to do this. The simplest: run your boat in a river. If this isn't possible, desalt your engine by using a hose with the special flushing fitting made for it, or by using the muffs that fit over the inlet grilles on the lower leg.

It is a good idea to adjust the water flow from the hose. It is correct when the tell-tale jet flows normally. If you are flushing your engine in a small tank (rubbish bin, old washtub, 200 litre drum, etc), you have to take certain precautions. For small engines without gears (2–3hp), remove the propeller to avoid accidents and water splashes. Otherwise, put the gears in neutral. After flushing, remove the fuel supply. Let the engine run at idle so as to empty all of the fuel lines and at the same time, spray a storage sealant into the carburettor(s) until the engine stops.

Empty the carburettor by removing the drain plugs.

◆ If your engine has been run a lot, it may be fouled. You can use an internal engine cleaner. While flushing the engine, spray a small amount of the cleaner a bit at a time into the carburettor(s) inlets for 5 minutes. Then, spray the cleaner into the inlets until the engine floods. Let the cleaner act for 20 minutes. Restart the engine and run in gear at full throttle for 5 minutes. This requires a special test tank. It is, of course, best to clean the inside of your engine when the boat is still on the water, before winterisation.

◆ When winterising your engine, it is critical to check the fuel and cooling systems function. If there is any doubt, investigate further. A weak tell-tale jet should alert you. A cloud of steam coming out of the engine exhaust is a sign of abnormally high engine temperature.

1 To start, run the engine for at least 30 minutes in fresh water. Here, a rubbish bin supplied with fresh water easily fits the bill.

2 Muffs placed on the water intakes allow thorough flushing of the outboard engine.

3 The tell-tale jet flow should be regular.

4 For greater safety, remove the propeller.

5 During flushing, the engine should idle without stalling or coughing.

6 Remember to drain the carburettor(s)...

7 ...via the drain plug to prevent deposits.

8 Also drain the fuel from the tank. The mixture loses combustibility after several months of storage.

9 If your engine has a lot of carbon deposits, don't hesitate to use an engine cleaner. This should be injected into the carburettor(s) after stalling the engine by starving it of fuel. Clean the engine head, then spray with penetrating oil.

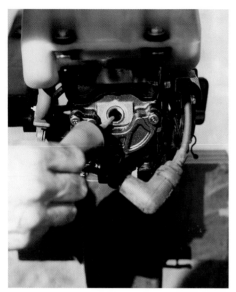

To complete the lubrication, an injection of two-stroke or storage oil into the cylinders is highly recommended.

Lubrication

Even though it isn't absolutely vital, it is a good idea to pour a little oil into the spark plug holes with an oil can to enhance the protection of the cylinders. A spoonful of oil in each cylinder should be sufficient.

With the kill cord removed or the spark plugs earthed, pull on the starter cord or push the starter button so that oil coats the cylinder lining. Replace the spark plugs.

Fuel filter

If your engine has a fuel filter, unscrew it and clean it with petrol. If it is too clogged or damaged, it must be replaced.

Draining the gearbox

1 Note the screws: the vent hole screw and the drain screw.

2 Drain oil in the lower leg into a clean container so that you can spot any traces of water.

Draining the gearbox

At the end of each season, or every 100 hours, the gearbox oil should be changed.

Unscrew the upper vent screw (marked 'oil level'), located under the anti-cavitation plate. Remove the lower drain screw. Let the oil drain into a small container.

Make sure that the drained oil is not emulsified. If it is, then water is getting into the gearbox. Check the water-tightness of the vent and drain screws. If the seals and seats seem in good condition, there are two other possible points of entry for water. One is the propeller shaft, the other is the drive shaft under the water pump.

Changing these seals is a job that requires special tools. You should therefore have the lower leg checked by an authorised dealer.

Refill the box through the lower fill hole with a tube of gearbox oil. Squeeze until the oil overflows through the upper vent hole. Plug the upper vent. With one hand, quickly remove the tube and, with the other, have the second screw ready to close the lower hole. Firmly tighten the screws.

Draining the gearbox

3 Make sure that the vent screw is removed before filling the gearbox with oil. The box is refilled through the lower hole with the engine in a vertical position.

4 Replace the upper screw, then the lower screw. Clean.

Cleaning and greasing

Remove the cover and clean the engine head with a brush and petrol. Blow or wipe dry.

Grease the various joints, then touch up the paint with the original colour. Complete the protection by spraying the entire engine head with a penetrating oil such as WD40.

Now clean the entire engine with a detergent to remove all traces of salt or grease.

Thoroughly rinse the engine with a jet of water or a pressure washer.

1 Grease the joints with a brush and marine grease.

2 Spray the entire engine with detergent. Clean with a brush.

3 Be sure to thoroughly rinse the engine with a jet of fresh water. Dry it until every trace of moisture is gone.

Blow the engine dry with compressed air or let it dry in the sun until every trace of moisture is gone.

Spray the entire engine with penetrating oil.

◆ When cleaning polyester engine covers and other plastic parts, never use paint thinner or acetone; use only detergent and water.
◆ Clean the threads on the transom clamp screws with a soft brush. Then grease the threads with a brush. Run the screws in and out several times.
◆ Grease all the engine joints by following the maintenance guide. To do this, you will need a grease gun sold by the dealer.

4 Spray the entire engine head with an external protection oil or penetrating oil.

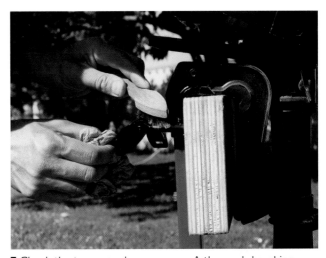

5 Check the transom clamp screws. A thorough brushing, followed by cleaning with petrol, will prevent any seizing.

6 Grease applied with a brush to the transom clamps will protect them and ensure that they work properly for the entire season.

7 The engine has several grease points that must be lubricated using a grease gun and cartridge sold by your engine dealer.

The propeller

Remove the propeller by either removing the split pin that secures the propeller end cap or castle nut, or by removing the nut and tab washer. Check the condition of the propeller blades. On small engines, the propeller will have a shear pin. When you remove it, it should come out easily, without using a pin punch. It shouldn't be bent or have any marks. Change it if there is any doubt as to its condition. Clean the propeller shaft. Make sure there aren't any fishing lines wrapped around it. Clean, then grease the propeller shaft. Remount the propeller, if possible with a new shear pin and split pin.

1 Remove the propeller, check the condition of the blades. Make sure there is no fishing line wrapped around the propeller shaft.

2 Clean the propeller shaft then apply marine grease.

3 Before remounting the propeller, check the shear pin. If it is damaged, change it. Always have one or two spares.

The recoil starter

Carefully examine the pull cord to make sure it isn't frayed or about to break. Check the condition of the rewind spring. Spray with penetrating oil.

Checking the pull cord when winterising will prevent you from finding yourself in spring holding a handle and broken cord. Change it if it shows any sign of deterioration.

The sacrificial anodes

If the anodes are more than 50% eroded, replace them. If they have become corroded/powdery, remove them, then clean the anode surface with a wire brush. Remove any traces of grease, paint or oil; then scrape the mounting surface clean.

Some advice

Drain the fuel tank. The fuel mixture or petrol degrades with age. Put in the amount of oil you would need when next refilling it. (Make a note that you have done this.) Shake the tank so the oil coats the interior. Next clean the tank exterior with detergent to remove any traces of salt or grease. Rinse with a jet of water. Once dry, spray it with penetrating oil.

Inspect, then wash the fuel line assembly: connectors, hose, and priming bulb.

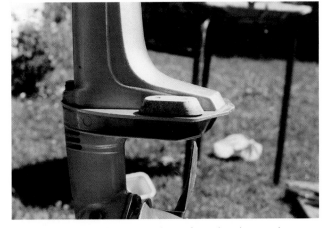

Check the anode. If it shows signs of erosion, it must be replaced. Otherwise, clean it thoroughly until you get a shiny surface.

Remember to degrease....

...then rinse the tank with a jet of water.

One more thing to check closely: the battery. Clean and recharge it before storing.

The battery

Modern batteries are reputed to be maintenance free. However, certain precautions still have to be taken before long storage to keep them in good condition.

First, remove the battery.

If your battery has caps, check and if necessary top up the electrolyte level with distilled water (1.5cm above the plates).

Second, recharge the battery. Make sure it is completely charged (consult the maintenance worksheet 'Check the battery charge').

Finally, clean the battery with plenty of water to dissolve the salt deposits. Wipe or dry. Store the battery in a dry place where it won't freeze.

Storing the engine

Store the engine in a vertical position in a dry, well ventilated place.

If you leave your engine on your boat's transom, consider protecting the power head with a cover. Don't use plastic covers.

Whew! It's finished. It is best to keep the engine in a vertical position in a dry, well ventilated place.

CONVERSION TABLE FOR COMMONLY USED UNITS OF MEASUREMENT

Length
1 inch = 25.4mm
1 foot = 12 inches = 304.8mm
1 yard = 3 feet = 914.4mm
1 mile = 1760 yards = 16093km

1 cm = 0.3937 inch
1 m = 1.0936 yards
1 km = 0.6214 miles

To convert miles to kilometres quickly, multiply by 8 and divide by 5.

1 international nautical mile = 1.852km

Liquid volume
1 gallon = 8 pints = 4 quarts
1 US gallon = 3.785 litres
1 UK gallon = 4.546 litres
1 UK gallon = 1.2 US gallons
1 US gallon = 0.83 UK gallons
1 litre = 0.264 US gallons
1 litre = 0.22 UK gallons

Capacity
1 cubic inch = 0.016387 litres
1 cubic inch = 16.387 cubic centimetres
1 litre = 61.024 cubic inches
1 cm^3 = 0.061 cubic inches

Weight
1 pound = 16 ounces = 0.4536kg
1 kg = 2.2046 pounds

Torque (tightening of screws)
1 in lb. = 0.115mg
1 ft lb. = 0.138mg

Temperature
32°F = 0°C
212°F = 100°C
C = 5/9 (F-32)
$F = \dfrac{9 \times C}{5} + 32$

Pressure
1 bar = 14.5037738 pound/square inch (absolute)
1 pound/square inch (absolute) = 0.0689476 bar

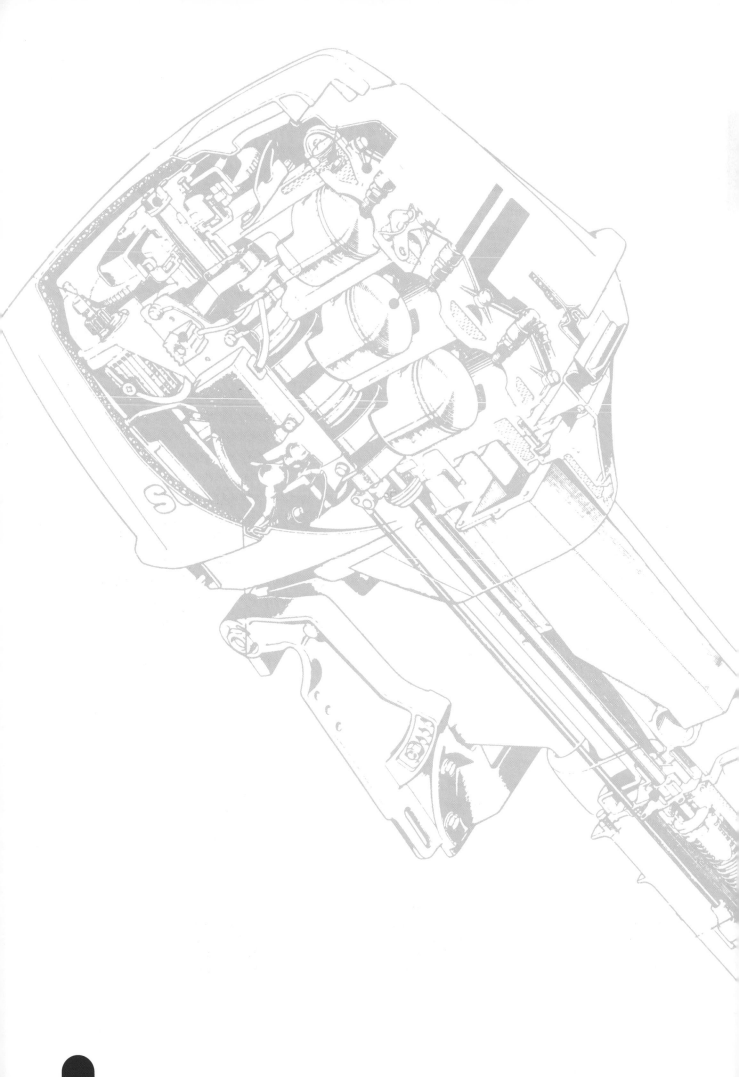

INDEX